Studies in Systems, Decision and Control

Volume 40

Series editor

Janusz Kacprzyk, Polish Academy of Sciences, Warsaw, Poland
e-mail: kacprzyk@ibspan.waw.pl

About this Series

The series "Studies in Systems, Decision and Control" (SSDC) covers both new developments and advances, as well as the state of the art, in the various areas of broadly perceived systems, decision making and control- quickly, up to date and with a high quality. The intent is to cover the theory, applications, and perspectives on the state of the art and future developments relevant to systems, decision making, control, complex processes and related areas, as embedded in the fields of engineering, computer science, physics, economics, social and life sciences, as well as the paradigms and methodologies behind them. The series contains monographs, textbooks, lecture notes and edited volumes in systems, decision making and control spanning the areas of Cyber-Physical Systems, Autonomous Systems, Sensor Networks, Control Systems, Energy Systems, Automotive Systems, Biological Systems, Vehicular Networking and Connected Vehicles, Aerospace Systems, Automation, Manufacturing, Smart Grids, Nonlinear Systems, Power Systems, Robotics, Social Systems, Economic Systems and other. Of particular value to both the contributors and the readership are the short publication timeframe and the world-wide distribution and exposure which enable both a wide and rapid dissemination of research output.

More information about this series at http://www.springer.com/series/13304

Hiram Eredín Ponce Espinosa
Editor

Nature-Inspired Computing for Control Systems

 Springer

Editor
Hiram Eredín Ponce Espinosa
Faculty of Engineering
Universidad Panamericana
Mexico City
Mexico

ISSN 2198-4182   ISSN 2198-4190 (electronic)
Studies in Systems, Decision and Control
ISBN 978-3-319-26228-4   ISBN 978-3-319-26230-7 (eBook)
DOI 10.1007/978-3-319-26230-7

Library of Congress Control Number: 2015955862

Springer Cham Heidelberg New York Dordrecht London

Printed on acid-free paper

Springer International Publishing AG Switzerland is part of Springer Science+Business Media
(www.springer.com)

*To Omar, for showing me on the path
of an undiscovered world*

Preface

Real-world control applications are difficult to tackle with conventional control systems because, in most of the cases, they present uncertainty, noise, and non-linearities. In that sense, intelligent control systems have been considered for designing robust and predictive controls, offering better performance than conventional control systems. However, new nature-inspired computing has proved to improve the performance over intelligent control systems, mainly using meta-heuristic optimization methods, social communication techniques, chemical structure-based learning algorithms, and so on.

Thus, the aim of this book is to present recent advances in nature-inspired computing applied for control systems. Readers with conventional control systems background will be motivated to undergo with these cutting-edge technologies, and advanced readers will be able to know and deeply understand a broad perspective of different methods and techniques.

At first, the book presents a review in the context of engineering control systems developed with nature-inspired computing. Then, the book is divided into three parts. The first part considers general approaches in control systems. Readers will find two contributions in terms of artificial neural networks, and improvements of techniques using hybrid methods. The second part considers on control tuning methods and adaptive control systems. This part summarizes some techniques and schemes that can be useful for real-world applications. For example, contributions show improvements to PID controllers using nature-inspired optimization, or develop fuzzy controllers in multi-input and multi-output systems. Other works present enhancements in neural controllers, or show advancements in neuro-fuzzy controllers. Lastly, the third part is dedicated to robotics systems using natural inspiration. Contributions in this part include controlling quadrotors using neuro-fuzzy techniques, developing algorithms for gait generation on legged robots, and designing new controllers for mobile robots using an optical flow approach. As shown, the book intends to give a practical overview in nature-inspired computing for engineering control applications.

In fact, this book cannot be possible without the support of many people. Thus, I want to thank Universidad Panamericana for its support in all concerns to this publication. I also want to acknowledge Dejanira Araiza-Illan and Roberto Ayala for their valuable comments and all the anonymous reviewers that helped me during the reviewing process. I also appreciate all the good words and advices from Omar Saldívar that motivated me through the development of this book. I acknowledge the assistance of Lourdes Martínez and Pedro Ponce. Finally, I want to thank all contributors for their amazing works and their research effort in the application of nature-inspired computing for control systems.

Mexico   Hiram Eredín Ponce Espinosa
August 2015

Contents

Part III   Robotics Applications

Contributors

José M. Araújo Grupo de Pesquisa em Sinais e Sistemas, Instituto Federal da Bahia, Salvador, BA, Brazil; Programa de Pós-Graduação em Engenharia Elétrica, Escola Politécnica da UFBA, Universidade Federal da Bahia, Salvador, BA, Brazil

José Roberto Ayala-Solares Department of Automatic Control and Systems Engineering, University of Sheffield, Sheffield, UK

Khaled Belarbi Faculty of Engineering, Mentoury University, Constantine, Algeria

Abdesselem Boulkroune LAJ, University of Jijel, Jijel, Algeria

Jorge Brieva Faculty of Engineering, Universidad Panamericana, Mexico City, Mexico

Isela Carrera Tecnologico de Monterrey, Mexico, Mexico

Israel Cayetano Tecnologico de Monterrey, Mexico, Mexico

Hachemi Chekireb LCP, ENP, Algiers, Algeria

Carlos E.T. Dórea Departamento de Engenharia de Computação e Automação, Universidade Federal do Rio Grande do Norte, 59078-900 Natal, RN, Brazil

Jose Gallardo Tecnologico de Monterrey, Mexico, Mexico

Mircea Hulea Faculty of Automatic Control and Computer Engineering, Department of Computer Science and Engineering, Gheorghe Asachi Technical University of Iasi, Iasi, Romania

Andreea Ion Department of Automatic Control and Systems Engineering, University Politehnica of Bucharest, Bucharest, Romania

Salim Issaouni LCP, ENP, Algiers, Algeria

Arturo Molina Tecnologico de Monterrey, Mexico, Mexico

Ernesto Moya-Albor Faculty of Engineering, Universidad Panamericana, Mexico City, Mexico

Monica Patrascu Department of Automatic Control and Systems Engineering, University Politehnica of Bucharest, Bucharest, Romania

Hiram Eredín Ponce Espinosa Faculty of Engineering, Universidad Panamericana, Mexico City, Mexico

Pedro Ponce Tecnologico de Monterrey, Mexico, Mexico

Jose Rodriguez Tecnologico de Monterrey, Mexico, Mexico

Hugo Salcedo Tecnologico de Monterrey, Mexico, Mexico

Oscar A. Silva Innovación y Robótica Estudiantil UTFSM, Valparaíso, Chile

Miguel A. Solis Centro de Robótica UTFSM, Valparaíso, Chile

Nesrine Talbi Faculty of Sciences and Technology, Department of Electronic, Jijel University, Jijel, Algeria

Bo Xing Computational Intelligence, Robotics, and Cybernetics for Leveraging E-Future (CIRCLE), Faculty of Science and Agriculture, Department of Computer Science, School of Mathematical and Computer Sciences, University of Limpopo, Sovenga, Limpopo, South Africa

The Power of Natural Inspiration in Control Systems

Hiram Eredín Ponce Espinosa and José Roberto Ayala-Solares

Abstract Throughout history, nature has always been an inspiration for mankind. It is not an exaggeration to say that almost every human invention, from engineering to social sciences, has been an attempt to replicate nature. In fact, nature continues to play an important roll in different human activities. From a scientific perspective, nature-inspired methods have proven to be an efficient tool for tackling real-life problems that are difficult to solve because of their high complexity or the limitation of resources to analyze them. The core idea is the fact that several natural phenomena, from simple to complex, always try to optimize certain parameters. Thus, this chapter gives an overview of nature-inspired methods from computational point of view, and summarizes key contributions of this book that focuses on methods that can simulate natural phenomena using computers, and the benefits of applying this methodology to the analysis and design of engineering control systems.

1 Naturally Inspired Methods

Throughout history, nature has always been an inspiration for mankind. It is not an exaggeration to say that almost every human invention, from engineering to social sciences, has been an attempt to replicate nature. In fact, nature continues to play an important roll in different human activities. From a scientific perspective, nature-inspired methods have proven to be an efficient tool for tackling real-life

H.E. Ponce Espinosa (✉)
Faculty of Engineering, Universidad Panamericana, 03920 Mexico City, Mexico
e-mail: hponce@up.edu.mx

J.R. Ayala-Solares
Department of Automatic Control and Systems Engineering, University of Sheffield, Sheffield S10 2TN, UK
e-mail: jrayalasolares1@sheffield.ac.uk

© Springer International Publishing Switzerland 2016
H.E. Ponce Espinosa (ed.), *Nature-Inspired Computing for Control Systems*,
Studies in Systems, Decision and Control 40, DOI 10.1007/978-3-319-26230-7_1

problems that are difficult to solve because of their high complexity or the limitation of resources to analyze them.

Computer science and Natural sciences have evolved considerably during the last years, and their interaction has produced new notions, techniques and methodologies that can be adapted to the complexities of real-life problems [12].

The last years of the 20th century mark the birth of natural computing, defined as a research field that focuses on the computation that takes place in nature. Natural computing works with three methodologies [6, 12]:

1. Human-designed problem-solving techniques inspired by nature.
2. Synthesis of natural phenomena based on computer simulations.
3. Use of nature-inspired materials to perform computations.

The core idea is the fact that several natural phenomena, from simple to complex, always try to optimize certain parameters. For example, ants can find the shortest path to a source food after several iterations, light always follows the path that takes the shortest time to travel, unstable chemical components interact with their environment until they reach stability, the hexagonal honeycomb shape in beehives maximizes the quantity of honey stored, while using the less quantity of wax in its construction. These few examples can provide the reader with a glimpse of why nature has inspired scientists in several research areas.

Throughout this book, several works about the advantages of nature-inspired methods are considered. In particular, this book focuses on methods that can simulate natural phenomena using computers, and the benefits of applying this methodology to the analysis and design of engineering control systems.

2   Types of Naturally Inspired Methods

The list of nature-inspired methods is vast and continues expanding [2, 3, 8, 12]. Several authors have proposed different classifications [2, 3], but in general, nature-inspired methods can come from three different areas of inspirations: biological, chemical, and physical [3]. Following, these naturally inspired methods are described, and Table 1 summarizes them.

2.1   Biological Inspiration

Biological processes have proven to be a great source of inspiration for nature-inspired methods, which can be divided in four categories: evolutionary algorithms, swarm algorithms, ecology algorithms, and reasoning algorithms. Following, these categories are briefly described.

Table 1 Comparative chart of different nature-inspired methods

Algorithm	Inspiration	Advantages	Disadvantages
Genetic algorithm (GA)	Micro-level genetic adaptation of organisms	They work well with problems whose search space is huge and there is no mathematical expression that can fully describe them	There are several parameters that need to be tuned and most of the time these are based on heuristics. In general, these algorithms tend to converge to local optima
Genetic programming (GP)	Micro-level genetic adaptation of organisms		
Evolution strategies (ES)	Theory of adaptation and evolution by means of natural selection		
Differential evolution (DE)	Micro-level genetic adaptation of organisms		
Evolutionary programming (EP)	Macro-level evolution of organisms		
Grammatical evolution (GE)	The generation process of a protein from genetic material		
Gene expression programming (GEP)	The replication and expression process of the DNA molecule		
Memetic algorithms (MA)	The interaction of genetic and cultural evolution		
Particle swarm optimization (PSO)	The social behavior of birds looking for food		
Ant colony system (ACS)	The ability of ants to find the shortest path between their nest and the food source	Compared with evolutionary algorithms, these can escape from local optima more easily. Several improvements allow these algorithms to work efficiently in high-dimensional problems	The implementation and usage of several agents can make these algorithms memory demanding and computationally expensive. Sometimes it may be necessary to parallelize the algorithm
Bees algorithm (BA)	The hierarchical structures within beehives		
Bacterial foraging optimization (BFO)	The processes and mechanisms that occur within bacterial populations		
Artificial immune system (AIS)	The clonal selection principle		
Intelligent water drops (IWD)	The processes that occur within natural river systems		
PS2O algorithm	Extension of the original PSO to consider the notion of symbiotic co-evolution between species	Good balance between exploration and exploitation	Given the complexity of the different attributes they are simulating, these types of algorithms can be computationally expensive
Invasive weed optimization (IWO)	The ecological process of weed colonization and distribution		
Biogeography-based optimization (BBO)	The study of immigration and emigration of species across time and space		

(continued)

Table 1 (continued)

Algorithm	Inspiration	Advantages	Disadvantages
Artificial neural networks (ANN)	Mimics the structure and function of neurons in the brain	These are considered as universal function approximators that can identify insightful patterns within datasets. Artificial Neural Networks have been applied successfully in different categories, which include regression and classification problems, data and image processing, system identification and control, among others	They are considered as black boxes that map inputs to outputs without providing a full description of the dynamics that connect these two. Furthermore, the training phase requires the tuning of several parameters and a large diversity of examples is required so that the network can learn meaningful patterns
Fuzzy logic (FL)	The vagueness of human language and logic	Fuzzy Logic works extremely well with vague data and the results produced are transparent compared with the black-box approach of neural networks. Fuzzy Logic has been applied successfully in several new algorithms and products	There is no unified approach that can prove the stability of fuzzy models. Furthermore, interpretability is possible as long as the system is simple, or it is divided into subsystems
Artificial organic networks (AON)	The characteristics of organic compounds in chemistry	These algorithms can deal with a variety of problems, and statistically, they can find global optima	In general, most of the algorithms tend to be slow given the complexity of the elements that they are simulating
Molecular computing (MC)	Computations based on molecules		
Chemical reaction optimization (CRO)	The natural process of converting unstable substances into stable ones		
DNA computing	The properties and characteristics of DNA		
Simulated annealing (SA)	The annealing process in metallurgy		
Extremal optimization (EO)	The field of statistical physics		
Harmony search (HS)	The improvisation skills of Jazz musicians		
Cultural algorithm (CA)	The principles of cultural evolution		
Quantum computing (QC)	The concepts and properties of quantum mechanics		

2.1.1 Evolutionary Algorithms

Evolutionary algorithms are inspired by the processes and mechanisms of biological evolution [2, 3, 8]. Some of the algorithms that belong to this category are:

- Genetic Algorithm (GA): population-based stochastic search algorithms inspired on the micro-level genetic adaptation of organisms.
- Genetic Programming (GP): similar to genetic algorithms, although the difference resides in the tree-like representation of the solution.
- Evolution Strategies (ES): optimization algorithms based on the theory of adaptation and evolution by means of natural selection.
- Differential Evolution (DE): similar to genetic algorithms, but here the mutation operator differs in that it is the result of an arithmetic operation between individuals.
- Evolutionary Programming (EP): similar to genetic algorithms, although this technique focuses on the macro-level of evolution.
- Grammatical Evolution (GE): optimization technique inspired on the generation process of a protein from genetic material.
- Gene Expression Programming (GEP): this algorithm is inspired by the replication and expression process of the DNA molecule.
- Memetic Algorithms (MA): these algorithms are inspired by the interaction of genetic and cultural evolution.

In general, evolutionary algorithms work well with problems whose search space is huge and there is no mathematical expression that can fully describe them. However, the main disadvantage is that these types of algorithms tend to converge to local optima.

2.1.2 Swarm Algorithms

The inspiration for swarm algorithms comes from the collective intelligence that emerges when a large number of homogeneous agents cooperate for a certain goal in the environment [2]. Some of the algorithms that possess this characteristic are:

- Particle Swarm Optimization (PSO): population- and trajectory-based stochastic search algorithms inspired by the social behavior of birds looking for food.
- Ant Colony System (ACS): similar to PSO, this is a population- and trajectory-based stochastic search algorithm that is inspired by the ability of ants to find the shortest path between their nest and the food source.
- Bees Algorithm (BA): optimization technique inspired by the hierarchical structures within beehives.
- Bacterial Foraging Optimization (BFO): algorithm inspired by the processes and mechanisms that occur within bacterial populations.

- Artificial Immune System (AIS): population-based algorithm based on the clonal selection principle.
- Intelligent Water Drops (IWD): population-based algorithm based on the processes that occur within natural river systems.

The main advantage of swarm algorithms is that they can escape from local optima more easily than evolutionary algorithms. However, the implementation and usage of several agents can make these algorithms memory demanding and computationally expensive.

2.1.3 Ecology Algorithms

Ecology algorithms are inspired by the interaction of living organisms with the environment [2]. Some of the algorithms that belong to this category are:

- PS2O Algorithm: optimization algorithm that extends the original PSO to consider the notion of symbiotic co-evolution between species.
- Invasive Weed Optimization (IWO): this is a stochastic search algorithm based on the ecological process of weed colonization and distribution.
- Biogeography-Based Optimization (BBO): this algorithm is inspired by the study of immigration and emigration of species across time and space.

These algorithms share several characteristics with swarm algorithms. In particular, they have a good balance between exploration and exploitation. However, these types of algorithms can be computationally expensive given the complexity of the different attributes they are simulating.

2.1.4 Reasoning Algorithms

The inspiration for these types of algorithms comes from the ability of life beings to process vague information from the environment and still produce an acceptable response depending on the input [3]. Two of the most popular schemes are:

- Artificial Neural Networks (ANN): this is a whole research field that mimics the structure and function of neurons in the brain. It focuses on developing computational models with a network-like structure that can be used to solve a great variety of scientific and engineering problems.
- Fuzzy Logic (FL): a research area inspired by the vagueness of human language and logic. Its application has produced successful results from control theory to artificial intelligence.

The main advantage of artificial neural networks and fuzzy logic is the ability to identify hidden patterns within complex data structures [7]. Once the patterns have been found, predictions can be performed fast and easily. However, the biggest disadvantage is that the training process for the neural network or the fuzzy model is not trivial and there are several parameters that need to be tuned to avoid the overfitting problem. To overcome this, several neuro-fuzzy hybridization techniques have been developed that try to combine the strengths of these two research areas.

2.2 Chemical Inspiration

As the name suggests it, these kinds of nature-inspired methods are based on chemical processes [5, 9]. Some algorithms that have chemical inspiration are:

- Artificial Organic Networks (AON): artificial intelligence technique inspired by the characteristics of organic compounds in chemistry.
- Molecular Computing (MC): computational approach that performs computations based on molecules.
- Chemical Reaction Optimization (CRO): optimization technique inspired by the natural process of converting unstable substances into stable ones.
- DNA computing: computational technique that makes use of the properties of DNA to make mathematical computations.

These techniques claim that chemical energy can be used as the heuristic guide for modeling and optimization processes. Since, chemical energy tends to a stable state; then, these chemically inspired methods promote minimal resources and finding global optima. In particular to DNA computing, it exploits parallelism to find solutions to hard problems. Disadvantages in these techniques is memory consuming.

2.3 Physical Inspiration

There are several nature-inspired methods that have their foundation in physical processes like metallurgy, music, and complex dynamics systems [3]. Some nature-inspired methods that have physical inspiration are:

- Simulated Annealing (SA): optimization algorithm based on the annealing process in metallurgy.
- Extremal Optimization (EO): optimization technique inspired by the field of statistical physics.
- Harmony Search (HS): this is a stochastic optimization algorithm based on the improvisation skills of Jazz musicians.
- Cultural Algorithm (CA): optimization technique inspired by the principles of cultural evolution.
- Quantum Computing (QC): a new computational paradigm inspired by the concepts and properties of quantum mechanics.

The main advantages of these algorithms are that they can deal with a variety of problems, and statistically, they can find global optima. However, most of the algorithms tend to be slow. Furthermore, harmony search and cultural algorithms have been seen as special cases of evolutionary algorithms [1, 11, 14].

3 Unconventional Control Systems

The above description considers Nature as the primary resource for inspiration in different computational methods. Since there are several applications for these techniques, this book will be reserved for those applications in terms of engineering control systems.

In control theory, conventional control systems are techniques employed for designing systems with desired behaviors. It means that any variable of interest in a mechanism, machine or device is maintained or altered with a defined rule [4]. Modern control theory has been applied in real-world problems, using the information of the plant. However, real-world problems are ill of uncertainty and vagueness that alter the behavior of the plant [4, 13]. As a result, conventional control techniques are not able to handle systems properly. In that sense, novel unconventional control systems have been applied [4].

Unconventional control systems are those methods employed to counter real-world problems in control systems. Typical problems identified are [4, 9, 10]: nonlinearities in the plant, impossibility for modeling the plant, uncertainty and noise in the system, vagueness in the behavior of the plant, dynamic operational point, and so forth. In that sense, unconventional control systems compute the control signal using different approaches such as: intelligent control and nature-inspired computing. On the one hand, intelligent control systems have used recently applying techniques like fuzzy logic and artificial neural networks. Complexity of real-world systems forced to mix these methods creating hybrid neuro-fuzzy controllers [7]. On the other hand, naturally inspired methods have been applied for improving the performance of intelligent control systems [4, 13].

In fact, unconventional control systems are designed to minimize the uncertainty and vagueness of engineering systems, by using controllers with high autonomy and robustness. It means that unconventional control systems provide enough flexibility and freedom to control laws to learn, predict and adapt to actual circumstances in systems. This flexibility and autonomy for handling unexpected behaviors bring to unconventional control systems the opportunity to design controllers without having the model of the plant [4].

3.1 Challenges in Unconventional Control Systems

There are still many challenges when using unconventional control systems [4, 13]. For instance, intelligent control and nature-inspired methods require more computational power than conventional control systems. In addition, real-time controllers are also a problem because these methods are highly time consuming. Since, nature-inspired methods need additional information about the performance of the system, they also are memory consuming. These challenges are not easy to tackle, and many efforts are still required.

3.2   Trends in Nature-Inspired Control Systems

Throughout this book, different applications are reviewed in which nature-inspired techniques for control systems have been employed to improve the performance of them. Thus, the book is organized as follows:

Part I begins with general control approaches like the implementation of spiking neural networks in hardware, a type of artificial neural networks, as the biological inspiration in a real-time control for laser spot tracking; and it finishes with the design of a hybrid control system using artificial neural networks and a new proposal so-called grey wolf optimization.

Part II explores some recent works in control tuning and adaptive control systems. For example, evolutionary algorithms were tested to perform modeling and control of industrial plants using auto-tuning of proportional-integral-derivative (PID) controllers; a fuzzy controller in a multiple-input and multiple-output nonlinear systems was implemented in which hybrid elite genetic algorithms and Tabu search methods were employed for optimal tuning; a neural controller was tuned for a special input-output manifolds in constrained linear systems; and an adaptive neuro-fuzzy controller was designed for induction machines.

Lastly, Part III presents some robotics applications such as: the design of a neuro-fuzzy controller for quadrotors using an adaptive-network-based fuzzy inference system (ANFIS); a novel image processing approach for mobile robots using a real-time optical flow technique based on Hermite transform; and the implementation of evolutionary function approximation for gait performance on legged robots.

References

1. Ali, M.Z., Morghem, A., Albadarneh, J., Al-Gharaibeh, R., Suganthan, P., Reynolds, R.G.: Cultural algorithms applied to the evolution of robotic soccer team tactics: a novel perspective. In: Evolutionary Computation (CEC), 2014 IEEE Congress on, pp. 2180–2187. IEEE (2014)
2. Binitha, S., Sathya, S.S.: A survey of bio inspired optimization algorithms. Int. J. Soft Comput. Eng. 2(2), 137–151 (2012)
3. Brownlee, J.: Clever algorithms: nature-inspired programming recipes. Jason Brownlee (2011)
4. King, R.E.: Computational intelligence in control engineering. CRC Press, Boca Raton (1999)
5. Lam, A.Y.S., Li, V.O.K.: Chemical reaction optimization: a tutorial. Memetic Comput. 4(1), 3–17 (2012)

6. Martn-Palma, R.J., Lakhtakia, A.: Engineered biomimicry for harvesting solar energy: a bird's eye view. Int. J. Smart Nano Mater. **4**(2), 83–90 (2013)
7. Nasira, G.M., Kumar, S.A., Kiruba, M.S.: A comparative study of fuzzy logic with artificial neural networks algorithms in clustering. J Comput Appl **1**(4), 6 (2008)
8. Neumann, F., Witt, C.: Bioinspired computation in combinatorial optimization: algorithms and their computational complexity. In: Proceedings of the 14th Annual Conference Companion on Genetic and Evolutionary Computation, pp. 1035–1058. ACM (2012)
9. Ponce-Espinosa, H., Ponce-Cruz, P., Molina, A.: Artificial organic networks. In: Artificial Organic Networks, pp. 53–72. Springer, Berlin (2014)
10. Primerano, R.: A case study in system-level physics-based simulation of a biomimetic robot. Trans. Autom. Sci. Eng. **8**(3), 664–671 (2011)
11. Reynolds, RG.: An introduction to cultural algorithms. In: Proceedings of the third annual conference on evolutionary programming, pp. 131–139. Singapore (1994)
12. Rozenberg, G., Bck, T., Kok, J.N.: Handbook of natural computing. Springer Publishing Company, Incorporated, Berlin (2011)
13. Szederkenyi, G.: Intelligent control systems: an introduction with examples. Springer Science and Business Media, Berlin (2006)
14. Weyland, D.: A rigorous analysis of the harmony search algorithm: how the research community can be. Model. Anal. Appl. Metaheuristic Comput. Adv. Trends, 72 (2012)

Part I
General Control Approaches

Bioinspired Control Method Based on Spiking Neural Networks and SMA Actuator Wires for LASER Spot Tracking

Mircea Hulea

Abstract This chapter presents a new biologically inspired technique for automatically compensating the light spot deviation from the normal position for laser spot trackers. The method is based on hardware implementation of the spiking neural networks which provides fast response due to real time operation and ability to learn unsupervised when they are stimulated by concurrent events. For increasing the biological plausibility of the method, the spiking neural network controls the contraction of shape memory alloy (SMA) actuator wires that operates as the muscular fibres. These SMA wires are the most suitable actuators for being controlled by the electronic spiking neurons because the contraction force increases naturally with the spiking frequency. From our knowledge the laser spot tracking using spiking neural networks was not performed previously. Moreover, other original ideas represent the use of analogue implementation of the spiking neural networks for real time operation as well as the SMA actuator wires for more biological plausibility. To validate this method we implemented in hardware a spiking neural network structure that processes the input from a one dimensional photodiode array and controls a positioning system based on SMA actuator wires. The results show that the spiking neural network is able to detect the one-dimensional spot motion and to adapt the response time by Hebbian learning mechanisms to the spot wandering amplitude. Moreover, by driving two antagonistic SMA actuator wires the system is able to track the laser spot with low response time and acceptable precision. These results are encouraging to develop bio-inspired low power spot tracking system for enhancing the receiving accuracy in free space optical communications or for enhancing the efficacy of the photovoltaic systems. Moreover, the light tracking principle based on spiking neural networks and SMA wires can be successfully used in implementation of the light tracking mechanism of an artificial eye.

M. Hulea (✉)
Department of Computer Science and Engineering, Faculty of Automatic Control
and Computer Engineering, Gheorghe Asachi Technical University of Iasi,
Iasi, Romania
e-mail: mhulea@tuiasi.ro

© Springer International Publishing Switzerland 2016
H.E. Ponce Espinosa (ed.), *Nature-Inspired Computing for Control Systems*,
Studies in Systems, Decision and Control 40, DOI 10.1007/978-3-319-26230-7_2

13

Keywords Laser spot tracking · Spiking neural networks · Analogue bio-inspired neuron · Shape memory alloy actuator wires · Photosensitive panel

1   Introduction

The light spot tracking problem was solved in several approaches using artificial intelligence elements such as fuzzy neural networks [5, 8, 28] or multilayer perceptrons [32]. However, despite the multiple approaches of this problem, the use of spiking neural networks represents a new method for light spot tracking. The advantage of this type of neural networks represents their ability to model high complexity functions in a bio-inspired manner while having very low power consumption when implemented in analogue hardware.

The spiking neural networks are adaptable parallel structures that use time for information processing and learning [7, 27]. Typically, the processing unit of these networks is the neuron designed to mimic in the most plausible way the neuronal cell behavior [20]. Several neuron models of biological inspiration such as integrate-and-fire [10, 24] were developed to mimic the information processing elements which are the membrane potential and the detection of activation threshold. Other neuron models suitable for software implementation that mimic the different types of spiking behavior of the biological neurons were developed by Izhikevich [16, 17]. The simulation time of large biologically inspired neural networks can be reduced by designing computationally effective neuron models [9, 22, 23, 26] or by implementing dedicated computational systems [25, 29]. However, the lowest response time and power consumption of the neural networks is achieved when these are implemented in analogue hardware. Thus, the neural network that is used as processing unit for the tracking system presented in this chapter is based on an adaptable electronic neuron [11, 12]. We use this type of spiking neuron that was analyzed in [13] because it is implemented in low power analogue hardware and mimics accurately the natural mechanisms of Hebbian learning [2, 3, 30].

The spiking neural network should control a position mechanism that is able to move the sensor area in order to track the laser spot. The most common method to actuate mechanical parts for the tracking systems represents the brushless or stepper motors. In contrast, for our work we considered for moving the sensor panel the recently implemented and tested SMA actuator wires [4, 31]. This actuation method increases the biological plausibility of the tracking system because the SMA wires operates by contraction as the natural muscles and they are naturally controlled by the spiking neural network because contraction force increases with the spiking frequency generated by the neural network.

Despite the fact that the SMA actuator wires were used for actuation of a prosthetic hand [1] we test for the first time the ability of the spiking neural networks (SNN) to control the SMA wires [15]. Also, the ability of the SNN to detect the laser spot deviation was performed for the first time [14]. Moreover, this work represents the first attempt to use spiking neural networks and SMA actuator wires for light spot tracking.

2 Neural Network

For compensating the deviation from the normal position of the light receiver, the tracking device should receive the data from a photodiode array and control a positioning mechanism. The processing unit of the tracking device is a spiking neural network that is able to learn by association during light spot tracking. For this research, we implemented the analogue spiking neural network, the photoresistive area and the positioning system based on shape memory alloy (SMA) actuator wires.

Spiking neurons are the most plausible models of biological neurons because they accurately mimic the natural mechanisms of information processing and of adaptation. First category of processes includes temporal integration of the incoming excitatory and inhibitory stimulation, membrane potential, activation threshold and refractory period, while the main properties that give the neural network the ability to learn from previous experience are the short-term potentiation (STP) and long-term potentiation (LTP) of the synapses. These mechanisms as well as the posttetanic potentiation are modelled by the electronic neuron that represents the basic unit of the spiking neural network used for laser spot tracking. One of the advantages of the analogue implementation of the spiking neurons is that they are multiplexed in space obtaining the real time operation of the entire neural network.

2.1 The Learning Mechanism

The strength of a synapse or connection between two neurons is increased with a factor p each time the synapse is activated. A synapse is active when the presynaptic neuron sends information to the receiving or postsynaptic neuron. If two incoming synapses towards the same postsynaptic neuron are concurrently activated in a time window t_p then the strength of the synapses is increased using a factor q [2].Throughout this chapter we say that two events are concurrent if they take place in the time interval t_p. From the biological point of view, t_p is the duration of the short-term potentiation. The factor p models the posttetanic potentiation contribution to synaptic plasticity and q models the long-term potentiation rate implying that $p \gg q$.

For the neuron model used as the processing unit the weight variation associated with the posttetanic potentiation is described by Eq. (1), while for the long term potentiation the weight varies with the amount given by the Eq. (2).

$$\Delta w_p = V_1 \cdot (e^{-p}) \qquad (1)$$

$$\Delta w_q = V_2 \cdot (e^{-q}) \qquad (2)$$

where V_1 and V_2 are two constants given by the neuron model design. During the neuron idle state the synapses are depressed with a factor r whose value is significantly lower than q [18]. Therefore, if a trained synapse is not activated in a long period of time the synaptic weight will decrease to the minimum value following the law:

$$\Delta w_q = V_3 \cdot (1 - e^{-r}) \qquad (3)$$

where V_3 is a constant dependent on the neuron design.

2.2 The Light Sensing Panel

The network receives the input from two areas of photodiodes placed like in Fig. 1. The tracking area of sensors is used for light spot tracking while the learning area is used for adaptation of the network response to the spot deviation amplitude as well as for tracking.

   This structure allows the neural network to learn to associate the effect of the sensors in the learning area with the effect of neurons in the tracking area starting from the initial conditions presented in the sequel.

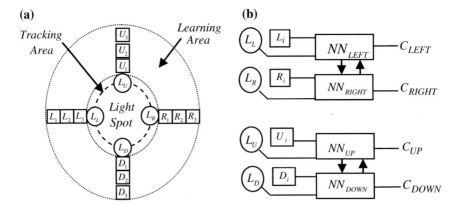

Fig. 1 a The panel of the sensors grouped in two areas—for tracking and respectively for learning. **b** The modules of the spiking neural network that generate the signals for controlling the positioning mechanisms of the sensor panel; The inhibitory connections shown by *arrows* are used when the spot covers two antagonistic tracking sensors simultaneously

In Fig. 1a the sensors from the tracking area L_L and L_R detect the spot movement on the horizontal axis, while the sensors L_U and L_D can be sensitive to the spot deviation on vertical axis. In the same manner the sensors L_i and R_i, $i = \overline{1,3}$ are sensitive to longer deviations of the light spot on the horizontal axis.

In order to make the network adaptable while being able the track the spot, the sensors from the tracking area are simultaneously active when the light spot is centered. For this system the spot size should cover half of the surface of the sensors from the tracking area when it is centered. In this setup small spot deviations determine covering of the whole surface of one sensor when the opposite sensor is not covered. The spot size was chosen for increasing the system detection accuracy because the input neurons from the tracking area activates when the spot moves with half of the sensor. Also, keeping the tracking neurons continuously active when the spot is centered inhibits the synaptic depression for the tracking neurons.

2.3 The Neural Network Structure

In order to control the positioning mechanism for compensating the bi-dimensional deviation of a light spot, the neural network should be divided in four principal modules that receive the inputs for each of the four directions—left, right, up and down. Figure 1b shows the modules of the neural network denoted by NN_{LEFT}, NN_{RIGHT}, NN_{UP}, and respectively NN_{DOWN} that are connected to the corresponding neurons from the tracking area (marked by circles) and from the learning area (marked by rectangles). The outputs of these modules that are designed to control the positioning mechanism for the photosensitive area are denoted respectively by C_{LEFT}, C_{RIGHT}, C_{UP}, and C_{DOWN}. Note that in order to compensate the light spot deviation from the central position the sensor panel should be moved in the same direction. Therefore, despite the fact that the position mechanism is placed behind the sensor panel that faces the laser implying that the motion on $left - right$ directions is reversed, for simplicity we will refer to the directions of the panel motions as seen from the laser side.

Taking into account that the structure of the neural network that detects the spot deviation on horizontal and respectively vertical axis are similar, for simplicity we implemented and analyzed the operation of the subsystem that tracks the spot on horizontal axis. Thus, the structure of the neural modules NN_{LEFT} and NN_{RIGHT} for controlling the $left$ and respectively $right$ directions on horizontal axis is presented in Fig. 2.

The neurons $N_{L[i]}$, and $N_{R[i]}$, $i = \overline{1,3}$ are the postsynaptic neurons for the sensors L_i and R_i, $i = \overline{1,3}$ that controls the outputs C_{LEFT} and C_{RIGHT} through the output neurons N_{CL} and N_{CR}, respectively. In the same way the sensors L_L and L_R are connected with the outputs C_{LEFT} and C_{RIGHT} like in Fig. 2. Taking into account that the centred spot covers two sensors with antagonistic effect on the positioning mechanism, the inhibitory neurons I_L and I_R are used for compensation of the

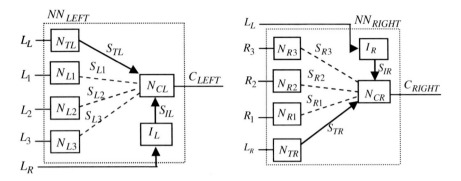

Fig. 2 The neural network structures that includes un-potentiated synapses that can be trained to compensate the spot displacement on *left* and *right* directions; The excitatory synapses S_{TL} and S_{TR}, as well as, the inhibitory synapses S_{IL} and S_{IR} that are connected to the training area are potentiated by PTP while the remaining synapses $S_{L[i]}$, and $S_{R[i]}$, $i = \overline{1,3}$ are potentiated mainly by LTP

excitatory activity of the tracking neurons N_{TL} and respectively, N_{TR} when the spot is centered. Therefore, when the spot moves to the *left*, the inhibitory activity of the neuron I_R is reduced determining less inhibition on the neuron N_{CL} that is stimulated by the excitatory activity of N_{TL}. Without inhibitory input activity the neuron N_{CL} starts to fire generating energy on the output C_{LEFT} that is used to move the sensory area to the left.

Note that while the spot is centred the two tracking neurons are activated simultaneously. The neural network structure was designed to remain active when the spot is centred in order to keep the synapses S_{TL} and S_{TR} potentiated by *PTP* to the maximum strengths without setting these strengths artificially. This structure allows the fastest response of the tracking neurons because of the maximum synaptic strengths. The structure of the neural modules NN_{UP} and NN_{DOWN} that control the sensor area motion on the vertical axis can be similar with the structure presented for the horizontal axis. Therefore, for compensating the receiver displacement on the *up* and *down* directions, the C_{UP} and C_{DOWN} outputs should be controlled by the corresponding photodiodes L_U and U_i respectively, L_D and D_i, $i = \overline{1,3}$ through similar synaptic configurations like in Fig. 2.

2.4 Neural Network Initial State

When the spot is not present on the sensor area and the network is not trained all synaptic strengths are minimum. After the laser is turned on and the spot covers the photodiodes L_L and L_R the synapses included by the corresponding neural paths are potentiated to the maximum values. For this work the maximum strengths of the synapses S_{TL} and S_{TR} that are activated by the sensors from the tracking area and the minimum strengths of the synapses $S_{L[i]}$ and $S_{R[i]}$, $i = \overline{1,3}$ corresponding to the learning area represents the initial synaptic configuration of the neural network. This configuration corresponds to the normal position of the spot when no spot deviations occurred previously. Thus, initial synaptic configuration of the neural network allows the sensors L_L and L_R to control a positioning device on the horizontal axis by activating accordingly the outputs C_{LEFT} and C_{RIGHT}, while the sensors L_i and R_i, $i = \overline{1,3}$ have no effect in controlling the positioning mechanism.

2.5 Neural Network Adaptation

The activation of the mechanisms that determine synaptic potentiation depend on the amplitude of the spot wandering. Therefore, if the spot deviation amplitude is higher than the tracking area and the spot reaches the learning area, the network starts learning by concurrent activation of the neural paths activated by the sensors from the tracking and respectively from the learning areas. For example, whether the light spot moves from the normal position to the left the L_L photodiode together with at least one of the photodiodes L_1, L_2, L_3 will stimulate simultaneously the neural network. Depending on the spot deviation amplitude the untrained synapses of the neural path between L_i, $i = \overline{1,3}$ and C_{LEFT} are strengthened. After several concurrent stimulations of the sensor L_L and the sensors from L group, the neurons $N_{L[i]}$, from the learning group will be able to activate the output C_{LEFT} in the same way as the tracking neuron N_{TL}. This orchestration of the tracking neurons activity increases the energy of the network response making the response of the tracking system faster. The active neurons for three spot deviation amplitudes when the synapses of the untrained paths are potentiated by *LTP* are shown in Fig. 3.

Thus, when the spot covers the sensor L_L and L_1 as in the Fig. 3a the untrained synapse S_{L1} is potentiated by the concurrent activity of the neurons N_{TL} that activates the postsynaptic neuron N_{CL} and the neuron N_{L1}.

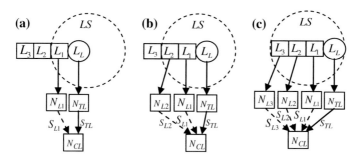

Fig. 3 Active synapses during the light spot (*LS*) deviation where the potentiated synapses are shown with *continuous arrows*; the un-potentiated synapses when the spot reaches **a** one, **b** two, or **c** three sensors from the learning area are shown with *dashed arrows*. These synapses are strengthened by *LTP* when the spot covers simultaneously at least one sensor corresponding to a trained neural path

Similarly, the synapses S_{L2} and S_{L3} are potentiated when the tracking neuron N_{TL} activates the output neuron N_{CL} as in Fig. 3b, and respectively in Fig. 3c. Therefore, the synapses marked with the dashed arrows are trained when the spot covers the tracking sensor L_L and the sensors L_i, $i = \overline{1,3}$ from the learning area. Taking into account that for *left–right* directions the network uses the same operation principles, the groups of photodiodes *UDL* and *R* will take progressively the role of L_L and respectively L_R during the training process. Therefore, the photodiodes placed in the tracking area act like supervisors for the activity of the photodiodes from the learning area.

On the other hand if the spot is static or the wandering amplitude is lower than the tracking area than the network starts learning at a lower rate by PTP. In this case the synaptic potentiation is determined only by presynaptic neuron activation. This mechanism is useful for potentiation of the synapses from the tracking area at the beginning of the experiment. After the synaptic strength S_{TL} increases above the threshold that makes it capable of activating N_{CL}, the neuron N_{TL} activation will strengthen the concurrent activated synapses, such as S_{L1}, S_{L2} and S_{L3}, depending on the spot deviation amplitude. Thus, if the spot wandering amplitude is high enough to cover the photodiodes L_1, L_2 and L_3 the corresponding input neuron N_{TL} will trigger long-term potentiation for the untrained synapses.

Due to the biological plausibility of the neuron model, the synapses that are not used are continuously depressed at a very low rate. Thus, whether the spot is steady in the normal position for a period of time that is longer than the duration of the complete synaptic depression, the learning process can start again from the initial conditions if the spot deviation amplitude increases. The synaptic depression ensures that the configuration of the synaptic strengths is suitable for compensating the spot wandering considering only the recent deviation history. This adaptation mechanism can reduce the unnecessary power that drives the SMA wires reducing panel oscillations and increasing the SMA wire lifetime.

2.6 The Positioning Mechanism for the Sensor Panel

The positioning mechanism of the system presented in this chapter is able to control the sensor area by moving it on the horizontal axis. The most common approach to implement the positioning systems represents the DC motors. Thus, the outputs of the neural network generate spikes whose frequency depends on the network activity. By integration of these spikes the analogue signals that can be decoded by a motor driver are obtained. The problem of controlling a DC motors speed using spikes was approached in [19] where the spiking neural network was designed in VLSI and implemented in FPGA.

However, the control of the DC motors using spiking neural networks implies the use of artificially designed circuit for conversion of the spiking activity of the

Table 1 Technical data for actuator wires used for actuating the sensor area

Diameter	Resistance	Pull force	Current for contraction	Stroke	Cooling time
0.006″	55 Ω/m	321 g	~410 mA	4 %	2 s

output neurons into the signals that rotates the DC motor. In contrast for implementation of our system we used SMA wires that are naturally driven by the spiking neural network because they actuate by contraction as the natural muscular fibres. This type of actuators brings more biological plausibility to the control system we proposed.

The outputs C_{LEFT} and C_{RIGHT} of the neural network generate spikes which frequency depends on the network activity. By integration of these spikes we obtained the characteristic of the network outputs as analogue signals that are amplified to power the actuator wires.

2.6.1 SMA Wires

The actuator wires are implemented with shape memory alloy (SMA) that contracts due to heating when they are electrically driven [31]. They are used mainly in robotics especially for hand prostheses because they are silent and model the muscular fibers behavior [1, 21]. In our experiments we used 0.006″ actuator wires type Flexinol LT. The technical data for the actuator wires is given in the Table 1 [6].

These wires are suitable to model the behaviour of natural muscles because they can act by contraction which intensity can be determined directly by the spikes frequency generated by the neurons. Thus, by increasing the spiking rate of the neural network output the current that drive the wires is increased bringing more contraction force to the SMA wires.

2.6.2 Positioning Mechanism Structure

The neural network generates continuous signals which energy is amplified to power the SMA wires. The schematic of the signal amplifier suitable for the system we implemented is shown in the Fig. 4.

Fig. 4 The power adaptation device (PAD) for driving the SMA actuator wires (AW) using the signal generated by the output neuron N_{CL}

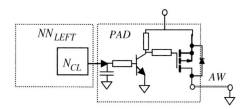

The neural network generates on each output spikes with different frequencies that depends on the number of input neurons activated simultaneously and on their spiking frequency. The role of power adaptation device (PAD) is to convert the spiking frequency into continuous analogue signals and to amplify the power of these signals for actuating the SMA wires. For example as shown in Fig. 4 the output of the NN_{LEFT} is integrated by a capacitor and the signal energy is amplified using the pair of transistors NPN-MOSFET that powers the actuator wire (AW).

Therefore, in order to implement a biologically plausible mechanism that can move the photosensitive area according to a light spot deviation from normal position we used SMA actuator wires because they are silent and operate as natural muscular fibres. These wires have limited stroke of about 4 % which implies that the wire length has to be significantly longer than the spot maximum deviation. In order to reduce the wire length while increasing the maximum spot wandering amplitude that can be compensated by the positioning system we implemented a lever mechanism as in Fig. 5.. The PA is mounted on a lever that is rotated by the actuator wires around a fixed point. Because for this system the force that is needed to move the PA is very low, we chose a much shorter lever for the SMA wires than the lever for PA. This ratio increases the PA motion amplitude and reduces the SMA wire stroke necessary to move the PA.

However, for this setup the PA follows a circular trajectory that might be unwanted if the spot deviates in a straight line. In order to be able to test the operation of the positioning device correctly we placed the laser module (LM) on a similar lever fixed at the same height as the lever of the positioning device. The lever length for the LM was adjusted to obtain the centred position of the light spot on the sensor panel.

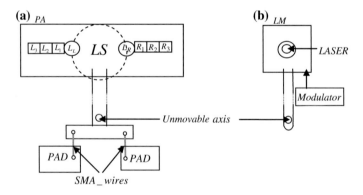

Fig. 5 **a** The structure of the positioning device for the photosensitive area (PA); the PA area is actuated by two SMA wires with antagonistic action controlled by two PDAs. For increasing the ratio between the PA motion amplitude and the SMA wire stroke we placed the PA on an lever that is able to rotate around a fixed point (FP); **b** the laser module (LM) used for testing the positioning device. The LM is mounted at the same height as the PA . For actuation of the PA's lever we used two PADs with antagonistic activity that were controlled by the neural networks NN_{LEFT} and respectively NN_{RIGHT}

3   Results

The experiments included three phases when we evaluated the spiking neural network response time, we test the adaptation abilities of the electronic synapses when the LASER spot moved across the senor area and we tested the ability of the SMA wires to move the photosensitive area to track the LASER spot when they are controlled by the electronic neural network.

The response time evaluation was performed using a microcontroller that simulated the activity of sensors when the spot deviates with constant speed across the sensors. For testing the adaptation abilities of the neural network that represents the second phase of the experiment the learning process is considered finished. In this case the photodiodes from the learning area are able to compensate the spot wandering as the sensors from the tracking zone. During the last phase of the experiment we evaluated the performance of a tracking device based on SMA wires controlled by the electronic neural network of biological inspiration when the spot was deviated deliberately from the centered position.

For all phases of the experiment one mono-synaptic neuron was connected to every sensor as presented in the Fig. 2.

3.1   Neural Network Response Time Evaluation

In order to test the neural network nonlinearity and learning rate when it is used for compensating the spot movement, we implemented a preliminary hardware device based on a microcontroller for generating the neural network input. Because the neural network modules have the same structure, in this section we simulated the activity of the photodiodes U_1, U_2 and U_3 that would detect the spot deviation towards *up*. This simulation is the same for all neural network modules that detect the spot wandering on the corresponding directions.

3.2   Spiking Output

The microcontroller was programmed to simulate the output of the photodiodes auxiliary circuit when PDs are covered by the light spot.

During this period the corresponding input neuron will generate action potentials like in Fig. 6. The spikes amplitude and duration depends on the synaptic weight and on the maximum energy generated by one impulse. For the maximum synaptic weights of the neural network used in this experiment, the spike duration is 60 μs and the amplitude is $V_s = 1.64\,\text{mV}$.

Considering that the spot size is constant, the moving velocity of the spot between the tracking sensors is proportional with the distance between the

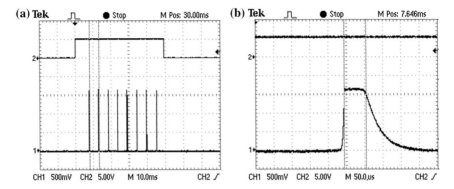

Fig. 6 The output of the neural network (*upper signal*); the microcontroller output modelling the activity of the U sensor (*lower signal*)

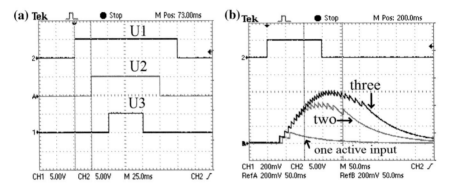

Fig. 7 **a** The simulation of the photodiode activity when the spot is moving across the sensors with constant velocity. **b** The difference between the network response when one, two and respectively three active network inputs

photodiodes d_I. The time T_I while the spot travels between two adjacent receivers decreases if the speed is higher. Thus, by controlling T_I and measuring the network output energy as well as the elapsed time until the network responds, it is possible to evaluate the network performance at different spot wandering velocities. The examples of the network output given in Fig. 7 were recorded for photodiode area edge of 3 mm and for $T_I = 25.2$ ms. Taking into account that $d_I = 9$ mm we obtain the spot wandering velocity $v = 0.36$ m/s for which we simulate the photodiode activity. Therefore, in order to assess the network ability to discriminate between the numbers of concurrently active sensors we recorded using an oscilloscope the neural network output.

The signals in Fig. 7a simulate the output of the sensors L_U, U_1 and U_2 if the spot passes across them at constant velocity from L_U to U_2 and backwards. The upper signal in Fig. 7b, represent the neural network input and the lower signal is the integrated response of the output neuron for three valid configurations of the active inputs. The presented results were obtained after the NN_{UP} neural network was trained meaning that each synapse was strengthened above the value that allows the presynaptic neurons to reach the postsynaptic action potential for the stimulated neurons. The waveform (b) allows measuring of the time lag of about 30 ms between the stimulus onset and the neural network response. Also, note the difference between the network output energy when the microcontroller simulates the activity of one, two or three sensors. The diagram shows that the neural network produces higher energy if the spot displacement is higher which means that the neural network acts as a regulator for compensating the light spot deviation.

The energy generated by the supervising neuron N_{LU} is maxim because this neuron used for tracking has to ensure that the light spot moving at normal speed remains centered when the synapses in the learning area are untrained. Also, the activity of these neurons increases the learning rate for the neurons that are connected to the photodiodes from the learning area.

3.3 Evaluation of the Network Adaptation Abilities

The role of the neurons in the learning area is to increase the nonlinearity of the neural network response in order to increase the device performance in spot tracking. For assessing the training rate of the neural network we evaluate the duration of the learning process by measuring the time t_L elapsed from the beginning of the network stimulation and the first spike generated by the output neuron N_{CD} that is the effect of the neuron activation. The initial weights of the synapses are zero meaning that the network is untrained.

The learning duration was measured when the synapses are potentiated by posttetanic potentiation (PTP) respectively by the short term and long-term potentiation (STP-LTP) mechanism.

The results presented in the Table 2 show that the learning period decreases with the number of active neurons for both learning mechanisms. The durations presented in the PTP column were obtained when the N_{LU} neuron was inactive. The learning rate significantly increases when LTP was triggered by the activity of the learning neuron N_{LU} due to the fact that the rate of the synaptic potentiation by STP-LTP is significantly higher.

Table 2 Learning duration

Active neurons (n_A)	Posttetanic potentiation PTP (s)	Long-term potentiation LTP (s)
1	2′52″	∼3
2	30″	∼3
3	20″	∼1

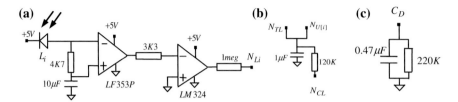

Fig. 8 **a** Light detecting circuit connected to the neural network inputs. **b** Auxiliary input circuit for neurons from hidden and output layers. **c** Circuit for integration of the network output energy

3.4 Laser Spot Movement Detection Experiment

The presence of the light spot on the photodiode surface is signalled to the corresponding neural network by the circuit presented in Fig. 8a. The photodiode resistance is converted by this circuit in a digital signal that is in high logic level when the light spot generated by a LASER type ADL-65055TL covers the photodiode type BPW20RF. The LASER spot diameter was adjusted using lens type CAY046 and for this phase of the experiment the LASER current was 30 mA.

Figure 9 presents the results obtained after the neural network activity was evaluated using the input generated by the photodiode area. The signal diagrams (a) and (b) presents the neural network output integrated using the circuit presented in Fig. 8c when the same test was performed in two different days.

In this case when the LASER spot moved across the photodiodes, the behaviour of the neural network was similar with that obtained when the network was stimulated by the microcontroller. The number of active neurons from the learning area determines the power of the network output and the response time.

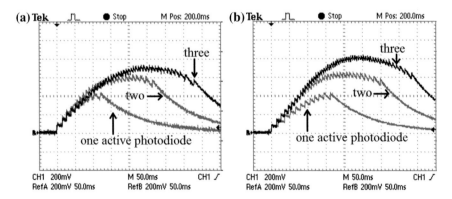

Fig. 9 Two network responses when the laser beam covers simultaneously one, two and respectively three photodiodes. The recordings (**a**) and respectively, (**b**) were performed after the network training in two different days

Thus, as shown in Fig. 9 if more neurons from the learning area are active, the output energy is higher and the response time is slightly decreased. The energy generated by the neural network can be calculated using the spike duration and amplitude which are given by the synaptic weights.

Because more neurons were activated when the spot displacement from normal position was higher implies that the network is sensitive to the amplitude of the spot deviation. This aspect increases the performance of the tracking system because the system reaction is amplitude dependent.

4 Laser Spot Tracking Experiment

The most important phase of the experiment represents the evaluation of the ability of the SMA wires to actuate the positioning device in order to compensate the spot wandering on the sensor area. The main goal of this experimental phase was to determine if the SMA actuator wires can react sufficiently fast when the light spot deviates from the central position because SMA wires contracts due heating that may delay their response. For this experiment we used the structure of the neural network presented in the Fig. 2 that uses one excitatory neuron driven by each photodiode included in the corresponding signal detection circuit shown in the Fig. 8. Also, the neural network uses two inhibitory neurons that are activated by the antagonistic sensors L_L and L_R from the tracking area.

4.1 Experimental Arrangement

Figure 10 presents several pictures of the experimental setup for testing the performance of the laser spot tracking system.

The first picture (a) shows the optical arrangement that includes the laser module facing the photosensitive area that is moved on *left–right* directions, on a circular trajectory, actuated by SMA wires. The PA pictured in the Fig. 10b uses four photodiodes for detection of the spot deviation in each direction. The spot normal position is centered between the photodiodes from the tracking area. The length of the levers for the sensor area and respectively for the laser module is 17.5 cm and the distance between the laser and the photosensitive area is 15 cm. The photosensitive area is actuated by two SMA wires of 27 cm length that are able to compensate on each direction the maximum spot deviation amplitude of 5 cm as shown in the Fig. 10d. When the PA is in normal position the wires are not tensioned because each wire should allow the antagonistic wire to contract. Thus, as pictured in Fig. 10d when the sensor area is moved to one side by contraction of the

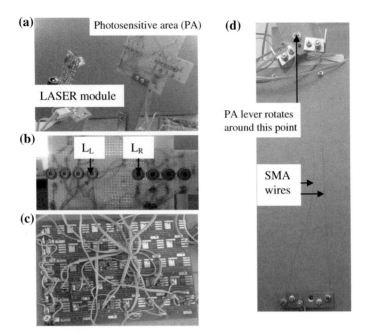

Fig. 10 Laboratory experimental setup for testing the performance of the tracking system based on spiking neural networks and actuator wires. **a** Optical arrangement. **b** Photosensitive area (PA). **c** PCB implementation of one network module (NN_{RIGHT}). **d** SMA actuator wires

corresponding wire, the antagonistic wire is relaxed and stretched. Taking into account that the wire maximum stroke is 4 % meaning that during contraction the 27 cm wire is shortened by ~1 cm, our experimental setup increases the wire contraction influence to 5 cm which represents a significant improvement of the ratio between the wire contraction and the maximum compensable deviation of the light spot.

The spot deviation from the normal position is detected by the electronic neural network. Figure 10c presents one module of the neural network implemented in PCB hardware that detects the spot deviation to the right. This neural network includes four presynaptic excitatory neurons and one postsynaptic neuron. The output power of the postsynaptic neuron is amplified to drive the actuator wires.

4.2 Tracking Performance Evaluation

The experimental results presented in the previous sections showed that the neural network responds sufficiently fast when the spot deviates and adapts its response to spot wandering amplitude. In this section we will determine the system response

Fig. 11 The spot is centered between the two tracking sensors; the tracking neurons are inactive

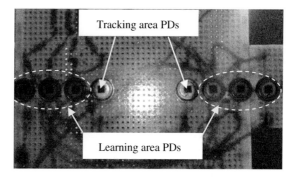

Tracking area PDs

Learning area PDs

time when it tracks the spot by controlling the position of the sensor panel. For this part of the experiment the synaptic strengths were set to the maximum values in order determine the fastest system response time. Also, because neural network adaptation facility is not used and the simultaneous activation of the tracking neurons from both sides of the spot is not necessary, the spot size was slightly reduced in order to keep all the input neurons silent when the spot is centered as in the Fig. 11.

The spot diameter can be estimated taking into account that the distance between the centers of the tracking PDs is 3.2 cm and the distance between the PDs from the learning area is 1 cm. The spot size ease the evaluation of the tracking system response time by monitoring only the neural network activity as presented in the sequel.

All results presented in this section were obtained by moving deliberately the laser lever. Before each test the lever was steady a period of time longer that 3 s that allows the SMA wires to cool down. Also the room temperature that influences the SMA wire relaxation state was around 28 °C.

In order to determine the system response time we use the laser diode (LD) type ADL-65075TR/L. For keeping active the inputs of the neural network we modulated the laser at 100 Hz using the external function generator G5100. The laser modulation was necessary because the auxiliary circuit for the photodiodes is designed to respond only to sudden changes in the received light intensity for reducing the ambient light influence on the neural network activity. This aspect determines continuous activation of the input neurons when the light spot covers the corresponding PDs. In order to obtain an acceptable optical power taking into account the spot diameter of 4 cm at 15 cm from the laser source, we set the current for the LD to 60 mA and connected a 40 pF capacitor in parallel with the LD.

Because for the experiments presented in the previous chapters the light was not modulated we started this phase of the experiment by preliminary evaluation of the neural network activity when the spot covers one or two spots. The activity of the spiking neural network was monitored using an oscilloscope type Tektronix TDS-210.

(a)   **(b)**

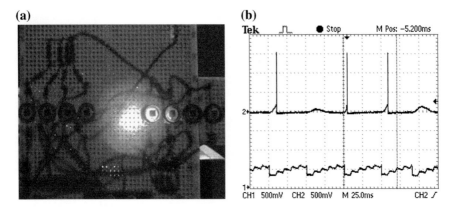

Fig. 12 **a** The laser spot covers the sensor from the tracking area. **b** The output of the postsynaptic neuron (*upper signal*) when activated by the tracking neuron whose input represents the *lower signal*

Figure 12 presents the activity of the neural module NN_{RIGHT} when the spot covers the photodiode L_R from the tracking area as in Fig. 12a. In order to be able to evaluate the neural network activity and to record the signals, during this preliminary test the power for the SMA wires was disabled. Figure 12b shows the activity of the output neuron N_{CR} when it is activated by the presynaptic neuron which input potential is shown by the lower signal. During normal operation the PAD integrates and amplifies the N_{CR} activity.

Similarly, Fig. 13a shows the picture of the photosensitive area when the spot covers two sensors—L_R from the tracking area and R_1 from the learning area. The waveform from Fig. 13b presents the output of the N_{CR} (upper signal) and the input potential of the presynaptic neuron N_{R1} that activates N_{CR}.

(a)   **(b)**

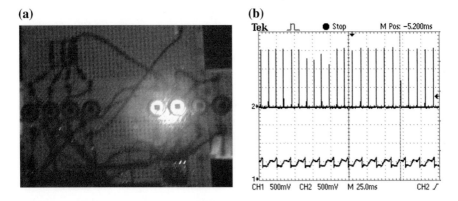

Fig. 13 **a** The light spot covers one sensor from the tracking area and one sensor from the learning area. **b** The *upper signal* shows the activity of the output neuron and the *lower signal* represents the input potential variation for the learning neuron L_{R1}

Note that the output neuron which activity is shown in Fig. 12b fires at maximum frequency of 18 Hz while in Fig. 13b the spiking frequency of the output neuron is about 100 Hz. As supposed, the spiking frequency of the network output is significantly higher when the light spot covers simultaneously two sensors than only one. This fact implies that the neural network is able to clearly discriminate between spot deviation amplitudes which increase the network sensitivity to the spot wandering amplitude.

In order to evaluate the ability of the tracking system to control the position of the photosensitive area for keeping the laser spot centered, we evaluated the response time and the tracking accuracy of the system.

4.2.1 Tracking System Response Time

In order to determine the period of time while the spot deviation is compensated we measured the duration of the tracking neurons activity that are stimulated directly by the photodiodes auxiliary circuit. Note that the tracking neurons are active only when the light spot stimulates the corresponding photodiode from the tracking area implying that when the spot is centered the neural network is silent. The duration of the neurons activity was evaluated based on the waveforms recorded on the oscilloscope.

Several results obtained when the deviated laser spot covers two sensors for each direction are shown in Fig. 14. These results are obtained when the maximum current that drives the SMA wires was limited to 300 mA representing 73 % of the maximum current mentioned by the wires manufacturer shown in the Table 2.

Fig. 14 shows the tracking neurons response when the laser spot deviation is compensated. In this case the wire that will contract is fully stretched and relaxed.

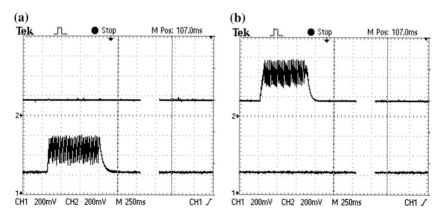

Fig. 14 The tracking neurons activity for evaluation of the response time t_L and t_R of the tracking system when the spot deviates towards the **a** *left* or **b** *right*; Before contraction the SMA wire is fully stretched and relaxed; The maximum driving current for the SMA wires is $I_A = 300\,\text{mA}$

Table 3 Response time for the tracking system (t_L and t_R)

Initial state of the wire	Fully stretched		Normal	
Maximum current for SMA wires (mA)	t_L (ms)	t_R (ms)	t_L (ms)	t_R (ms)
$I_A = 300$	813	766	1033	1160
$I_A = 350$	713	846	846	886
$I_A = 400$	560	506	700	706

This configuration is achieved, for example, by placing the panel to the maximum right position with the spot centered.

Thus, in the Fig. 14a the spot deviation to the left is compensated in the period t_L. The upper signal represents the input of the tracking neuron N_{TR} and the lower signal represents the input activity of N_{TL}. Similarly, the spot deviation to the right is compensated in the time t_R as shown in Fig. 14b. The gaps that appears on both signals are introduced by the oscilloscope when the time per division is greater than 100 ms. For having a qualitative view of the network response time we can estimate that $t_L = 700$ ms and $t_R = 630$ ms taking into account that one vertical division on the oscilloscope represents $M = 250$ ms.

Several values of the measured response time of the tracking system are given in the Table 3.

Each value represents the average of three time measurements performed using the oscilloscope cursors that were placed at the beginning of the first stimulation of the neuron input and respectively at the last neuron activation. The first two columns of the Table 3 present the response time for left and respectively right deviation of the sensor panel when the SMA wire was stretched before contraction. The last two columns present the values for the same parameters when the sensor panel lever was vertical before SMA wire contraction. On each row we present the response time for several values of the maximum current that drives the actuator wires.

4.2.2 Tracking Precision

The normal position of the light spot is centered when the sensor panel lever is vertical. Therefore, the system should be able to track the spot starting from this position implying that the SMA actuator wires are not stretched when they are relaxed. This may add a delay in the system response time because the panel is actuated after a period of time elapsed before the wire is tensioned. Indeed, in this case the network response time increases to $t_{OSC(L)} = 1.150$ s for left deviation and to $t_{OSC(R)} = 1$ s for right deviation as shown in the Fig. 15a, b. Note that for the deviations in both directions the opposite tracking neuron activates showing that the sensor panel deviates in the opposite direction determining the activation of the opposite photodiode from the tracking area. Therefore when the SMA wire is not

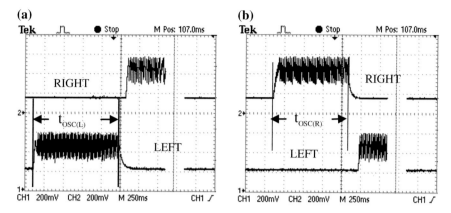

Fig. 15 The tracking neurons activity for evaluation of the response time of the tracking system when the spot deviates towards **a** *Right*. **b** *Left* starting from the normal position; The maximum SMA wire current is $I_A = 350$ mA

fully stretched the sensor panel is actuated sharply making the sensor area to oscillate due to inertial forces and slightly due to panel weight.

Several values of the $t_{OSC(L)}$ and $t_{OSC(R)}$ for $I_A \in \{300, 350, 400\}$ mA are given in the Table 4. These values were obtained using the same method that gives the data presented in the Table 3.

Our results showed that for $I_A = 300$ mA and $I_A = 350$ mA the oscillation of the sensor panel does not occur when the actuating wire is stretched before contraction.

For a qualitative illustration of the panel oscillation Fig. 16 presents two waveforms recorded from the oscilloscope that capture panel oscillations. The waveform (a) shows that the spot deviation to the left that is compensated in about *760 ms* when the sensor panel is pulled by the corresponding SMA wire. The wire contraction force determines the panel deviation in the opposite direction. The time difference t_i between the last activation of the input neuron N_{TL} and the first stimulation of N_{TR} gives us information about the panel velocity that in this case is estimated to $v_p = 0.4$ m/s. This value is in concordance with the value $v = 0.36$ m/s of the spot wandering simulated using the microcontroller (see Sect. 3.2). Similarly, t_f gives information about the panel velocity in the opposite direction. Note that

Table 4 Oscillation duration when inertial forces affects the sensor panel control

Initial state of the wire	Fully stretched		Normal	
Maximum current for SMA wires (mA)	$t_{OSC(L)}$ (s)	$t_{OSC(R)}$ (s)	$t_{OSC(L)}$ (s)	$t_{OSC(R)}$ (s)
$I_A = 300$	no osc.	no osc.	1.8	2.09
$I_A = 350$	no osc.	no osc.	1.4	1.89
$I_A = 400$	1.36	1.49	2.03	1.99

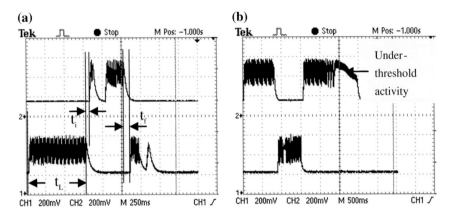

Fig. 16 Activity of the tracking neurons when the spot deviates from normal position (the panel lever is vertical) for $I_A = 400\,$mA

$t_f > t_i$ implying that the spiking neural network acts to reduce the oscillation amplitude.

However, the panel velocity that gives information about the inertial forces represents only estimation and is not the subject of the research presented in this chapter.

Similar behavior is observed in Fig. 16b that presents the tracking neurons activity when the spot deviates to the right starting from normal position. The unusual activity of the tracking neuron N_{TR} that is pointed by the arrow on the upper signal represents under-threshold stimulation of the neuron. This activity stopped because the panel is still moving slightly pulled by the actuator wire that is cooling down.

5 Discussions

The research presented in this chapter demonstrates the principle for mono-dimensional light spot tracking using a biologically plausible system. Despite the fact that we tested the laser spot tracking for horizontal axis, the same principle can be easily used to implement a control module that tracks the spot on vertical axis. Because the spot tracking on horizontal and respectively vertical axis is performed by independent modules that operates in parallel the combination of the two systems is able to perform bi-dimensional laser spot tracking.

In this experiment we limited the maximum current I_A that can drive the actuator wires to several values and determined the response time and regulatory performance of the control system. The results show that when the laser spot deviates, the tracking system respond for highest I_A in around 0.5 s to compensate the laser spot displacement. This implies that the SMA wires reacts unexpectedly fast when the light spot covers at least the diode from the tracking area taking into account that the wire contraction occurs due heating.

One of the interesting aspects is the fact that because the response of the actuator wires is fast, the inertial forces for the sensor panel occur. Due to these forces and to the contraction of the antagonistic actuator wire when the light reaches the opposite tracking sensor the panel tends to oscillate but it stabilizes after one or two oscillations.

However, our experiments showed that the cooling time of the wires, that according to the datasheet is 2 s, becomes a problem during fast oscillations of the laser spot. This problem could be easily solved using thinner wires for which the cooling time is as low as 0.15 s. Moreover, according to the technical characteristics of the SMA actuator wires among the cooling down methods the most effective is water with glycol [6].

Another behavior that affects the system tracking precision represents the panel area oscillations. These oscillations occur when I_A is near the maximum constructive value or when the wires are not stretched before contraction. This problem was solved partly by reducing the maximum current that drives the SMA wires with penalty in the system response time. Also, the oscillation problem could be solved by reducing the weight of the photosensitive panel.

6 Advantages of the Nature-Based Method for Control Systems

This chapter describes a biologically inspired method for laser spot tracking based on spiking neural networks and SMA actuator wires. The spiking neural network is able to track a laser spot by detecting the spot deviation from a central position using the output of two areas of photodiodes and by driving the photodiode panel using two antagonistic SMA actuator wires. The neural network is based on a neuron model that mimics the main mechanisms of the biological neurons regarding the information processing and the associative learning. The SMA wires mimic the behavior of muscular fibers because they act by contraction when they are stimulated by currents.

The spot tracking system presented in this chapter represents a high complexity control system because it is responsive to perturbations that vary in a continuous range and adapts continuously for better compensation of these perturbations. Despite the high complexity function of the control system, the design of the regulatory unit is simple because it uses spiking neurons implemented in analogue hardware that model intrinsically high complexity functions as the neural cells. Thus, this type of neurons allowed us to implement the processing unit of the tracking system for the horizontal axis using only ten excitatory neurons and two inhibitory neurons.

Moreover, the spiking neural network controls naturally the contraction of SMA actuator wires that reacts fast but smoothly as the natural muscular fibers. This behavior ensures a biological-like behavior of the tracking system that controls the position of a sensory area for keeping the light spot centered.

The advantages of the spiking neural network implemented in analogue hardware represent the real-time response because all neurons operate in parallel as well as the very low power consumption. For example, taking into account that one neuron with one synapse needs around 0.5 µW to operate, the power consumption of the whole neural network that performs the high complexity control function for spot tracking is about 6 µW. This value is significantly lower than the necessary power for any microcontroller that would implement even lower complexity control functions. Moreover, modeling higher complexity functions using algorithms increases the response time of the processing unit.

The control functions for the classical systems are designed to fulfill specific goals and when an unpredicted situation occurs the system may not be able to deal with the unknown perturbations. In contrast, the spiking neural network is adaptable being able to adjust the power that drives the actuator wires according to the previous perturbations. Moreover, when using spiking neural networks the exact model of the process is not necessary for designing the regulatory unit because spiking neurons act as regulators.

Considering all mentioned above the main advantages of the bio-inspired method for control system represents the adaptability, model-less design of the regulatory unit, very low power consumption and the real time response control unit. Moreover, using the proposed method the light tracking mechanism of the eye could be modeled.

7 Conclusions

The light spot tracking is a necessity for enhancing the quality of the link stability in FSO communications or for increasing the received light power in photovoltaic systems. Moreover, the most complex light tracking function is performed by the eye and the corresponding neural areas. Thus, the most suitable processing unit for solving this task can be the analogue implementation of the spiking neural networks due to highly non-linear behavior, fast response time and very low power consumption.

After testing the neural network performance by simulation of the light sensors activity using a microcontroller, the results showed that the velocity of the spot movement in strait line can reach easily 0.36 m/s while the consumed power of the active neurons is less than 6 µW. One important aspect represents the significant dependency of the network response energy on the number of neurons that are concurrently active. Similar network behavior was obtained when it was tested in real conditions using a laser spot that was moving across the photodiode area.

The compensable speed of the spot wandering using the spiking neural networks (SNN) makes SNN suitable for laser spot tracking in turbulent environment when used in free space optical communications.

However, the type of SMA actuator wires used in our experiment might be more suitable for compensation of the spot wandering in weak turbulence due to relatively high response time and very high cooling time.

The low power neural networks that control the actuator wires are suitable for improving the efficiency of the photovoltaic power supplies by continuously tracking the light spot of maximum intensity.

Considering the response time of our system and its smooth biological like behaviour the unidirectional light tracking principle demonstrated by our system is most suitable to be used in modelling and explaining the light tracking mechanisms of the eye. This mechanism for the light spot tracking could be the following: when the spot deviates from a central position on the retina the muscles contracts to rotate the eye for compensation of the spot displacement on the retinal sensorial area.

One of the future directions of our research will be based on this idea when we intend to model the light sensitive retinal cell using photodiodes and the muscles that rotate the eye using SMA actuator wires.

References

1. Andrianesis, K., Tzes, A.: Development and control of a multifunctional prosthetic hand with shape memory alloy actuators. J. Intell. Robot. Syst. **78**, 257–289 (2014)
2. Bi, G., Poo, M.: Synaptic modifications in cultured hippocampal neurons: dependence on spike timing, synaptic strength, and postsynaptic cell type. J. Neurosci. **18**, 10464–10472 (1998)
3. Dan, Y., Poo, M.: Spike timing dependent plasticity of neural circuits. Neuron **24**, 23–30 (2004)
4. Donmez, B., Ozkan, B., Kadioglu, S.: Precise position control using shape memory alloy wires. Turk. J. Elec. Eng. Comp. Sci. **18**, 899–912 (2010)
5. Duh, F., Lin, C.: Tracking a maneuvering target using neural fuzzy network. IEEE Trans. Syst. Man, Cybern. B, Cybern. **34**, 16–33 (2004)
6. Dynalloy Inc: Technical characteristics of Flexinol actuator wires (no year specified)
7. Gerstner, W., Kistler, W.: Spiking neurons models: single neurons, populations, plasticity. Cambridge University Press, Cambridge (2002)
8. Haibin, S., Jingjing, B.: Maximum power point tracking algorithm based on fuzzy neural networks for photovoltaic generation system. pp. 353–357 (2010)
9. Hines, M., Eichner, H., Schürmann, F.: Neuron spitting in compute-bound parallel network simulations enables runtime scaling with twice as many processors. J. Comput. Neurosci. **25**, 203–210 (2008)
10. Hopfield, J., Brody, C.: What is a moment? transient synchrony as a collective mechanism for spatiotemporal integration. Proc. Natl. Acad. Sci. USA **98**, 1282–1287 (2001)
11. Hulea, M.: A model of silicon neurons suitable for speech recognition. Control Eng. Appl. Inform. **10**, 32–41 (2008)
12. Hulea, M.: A new method to obtain non-volatile memory for networks of spiking neurons. Memoirs of Sci. Sect. **XXXIII**, 129–146 (2010)

13. Hulea, M.: The mathematical model of a biologically inspired electronic neuron for ease the design of spiking neural networks topology. pp. 282–287 (2011)
14. Hulea, M.: Using spiking neural networks for light spot tracking. In: 20th European Signal Processing Conference, pp. 1708–1712. Bucharest (2012)
15. Hulea, M., Caruntu, C.: Spiking neural network for controlling the artificial muscles of a humanoid robotic arm. In: International Conference on System Theory, Control and Computing, pp. 163–168. Sinaia (2014)
16. Izhikevich, E.: Which model to use for cortical spiking neurons? IEEE Trans. Neural Netw. **15**, 1063–1070 (2004)
17. Izhikevich, E.: Polychronization: computation with spikes. Neural Comput. **18**, 245–282 (2006)
18. Jacob, V., Brasier, D., Erchova, I., Feldman, D., Shulz, D.: Spike timing-dependent synaptic depression in the in vivo barrel cortex of the rat. J. Neurosci. **27**, 1271–1284 (2007)
19. Jimenez-Fernandez, A., Jimenez-Moreno, G., Linarea-Barranco, A., Dominguez-Morales, M., Paz-Vicente, R., Civit-Balcells, A.: A neuro-inspired spike-based PID motor controller for multi-motor robots with low cost FPGAs. Sensors **12**, 3831–3856 (2012)
20. Jolivet, R., Lewis, T., Gerstner, W.: Generalized integrate-and-fire models of neuronal activity approximate spike trains of a detailed model to a high degree of accuracy. Neurophysiology **92**, 959–976 (2004)
21. Joshi, P., Maass, W.: Movement generation with circuits of spiking neurons. Neur. Comput. **17**, 1715–1738 (2005)
22. Kobayashi, R., Tsubo, Y., Shinomoto, S.: Made-to-order spiking neuron model equipped with a multi-timescale adaptive threshold. Front. Comput. Neurosci. **3** (2009)
23. Lovelace, J., Cios, K.: A very simple spiking neuron model that allows for modeling of large, complex systems. Neural Comput. **20**, 65–90 (2008)
24. Maass, W., Bishop, C.: Pulsed neural networks. The MIT Press, Cambridge (1998)
25. Maass, W., Natschläger, T., Markram, H.: Computational models for generic cortical micro-circuits. Comput. Neurosci. A Compr. Approach pp. 575–605 (2004)
26. Markram, H., Shurmann, F.: Fully implicit parallel simulation of single neurons. J. Comput. Neurosci. **25**, 439–448 (2008)
27. OReylli, R., Munakata, Y.: Computational explorations in cognitive neuroscience. The MIT Press, Cambridge (2000)
28. Otieno, C., Nyakoe, G., Wekesa, C.: A neural fuzzy based maximum power point tracker for a photovoltaic system. In: AFRICON, pp. 1–6 (2009)
29. Rast, A., Mukaram Khan, M., Jin, X., Plana, L., Furber, S.: Universal abstract-time platform for real-time neural networks, pp. 2611–2618. IEEE Press, Piscataway (2009)
30. Swiercz, W., Cios, K.: A new synaptic plasticity rule for networks of spiking neurons. IEEE Trans. Neural Netw. **17**, 94–99 (2006)
31. Teh, Y.: Accurate force and position control of shape memory alloy actuators. Ph.D. thesis, The Australian National University (2008)
32. Wong, Y., Sundareshan, M.: Data fusion and tracking of complex target maneuvers with a simplex-trained neural network-based architecture, pp. 1024–1029. Bucharest (1998)

The Spread of Innovatory Nature Originated Metaheuristics in Robot Swarm Control for Smart Living Environments

Bo Xing

Abstract The main purpose of introducing ambient assistive living (AAL) robots is to assist the disabled and elderly people at home. In recent years, this field has evolved quickly because of the enormous increase in computing power and availability of the improved variety of sensors and actuators. However, design of AAL robots control system is a huge challenge, which require solving issues related to two classes: design of mechanical structure and development of an efficient control system. In this chapter, we focus on the latter topic, since even relatively low quality hardware can be used for solving sophisticated tasks if the software control it correctly. The chapter starts by giving a vision of what heterogeneous AAL robots is supposed to look like and how a human is to act, navigate and function in it. Particularly, we investigate the effect of artificial neural network (ANN) based control techniques for AAL robots. To enhance the accuracy and convergence rate of ANN, a new method of neural network training is explored, i.e., grey wolf optimization (GWO). Moreover, we provide an overview of applying emerging metaheuristic approaches to various smart robot control scenarios which, from the author's viewpoint, have a great influence on various AAL robot related activities, such as location identification, manipulation, communication, vision, learning, and docking capabilities. The findings of this work can provide a good source for someone who is interested in the research field of AAL robot control. Finally, we concludes with a discussion of some of the challenges that exist in the AAL robot control.

Keywords Ambient assisted living (AAL) · Robot control · Computational intelligence (CI) · Bacteria foraging optimization (BFO) · Bees algorithm (BA) · Glowworm swarm optimization (GSO) · Electromagnetism-like mechanism (EM) · Intelligent water drops (IWD) · Ambient assisted living (AAL) · Assistive technology services (ATSs) · Nature-inspired computing · Robot control · Metaheuristics

B. Xing (✉)
Computational Intelligence, Robotics, and Cybernetics for Leveraging E-Future (CIRCLE),
Faculty of Science and Agriculture, Department of Computer Science,
School of Mathematical and Computer Sciences, University of Limpopo, Private Bag X1106,
Sovenga, Limpopo 0727, South Africa
e-mail: bxing2009@gmail.com

© Springer International Publishing Switzerland 2016   39
H.E. Ponce Espinosa (ed.), *Nature-Inspired Computing for Control Systems*,
Studies in Systems, Decision and Control 40, DOI 10.1007/978-3-319-26230-7_3

1 Introduction

The ability to moving independently is essential for the full development of our lives. Traditionally, for people who have mobility in their upper limbs, the use of mechanical or electric wheelchairs represents a way to regain some of their mobile. However, it is important that the user must acquire knowledge and skills to handle them. In addition, the users have very limited options for take care of themselves. For these reasons, in recent years, there is an increasing interest in the development of assistive technology services (ATSs).

Generally speaking, ATSs aim at applying ambient intelligence technology to support independent living of people with disabilities and special demands [1]. From this perceptive, device-based (e.g., ambient assistive living robots, AAL robots for short) technologies can be seen as important tools that provide high standards to cognitive capabilities, autonomy and movement precision. Applications of AAL robots include independent living [2], surgery [3, 4], rehabilitation [5, 6], and entertainment [7], to name a few. As with any robotics, one of the most challenging problems in AAL robots (in particular with heterogeneous robot members) is the design of an appropriate control system, such as fusion of various sensor information, high accuracy actuation, and reliable software implementation. In this chapter, we intend to utilize artificial neural network (ANN) trained by a novel nature originated metaheuristics for fulfilling such control task.

With this in mind, the remainder of this chapter is organized as follows: First, Sects. 2 and 3 briefly introduce the relevant terms used throughout this study; Second, the focal problem of this chapter is elaborated in Sect. 4 which is followed by a detailed explanation of the employed methodology in Sect. 5; right after this, Sect. 6 describes the experimental environment setup of this chapter; Next the future work provided in Sect. 7 outlines several limitations of the present work; Finally, Sect. 8 draws the conclusion of this study.

2 Background of Ambient Assisted Living (AAL) Robot

According to [8], a robot can be broadly defined as an apparatus that is designed to act as an intelligent linkage between perceiving and acting. Under the umbrella of this definition, mobile robots are mainly referred to the robots the can relocate themselves from one location to another autonomously, i.e., without their surrounding human beings' interference. Unlike their industrial counterparts that can move only along a specific trajectory, mobile robots are capable of moving around freely within a predetermined workspace for the purpose of fulfilling a goal. Following this trend, AAL robot has become an active research in the area of healthcare.

The main purpose of introducing AAL robots is to assist the disabled and elderly people at home [9–11]. Nowadays, there are many types of AAL robots being developed in research laboratories and by companies all over the world. In general, they can be categorized into three categories: robots assisting with daily living activities (such as feeding and dressing), robots assisting with instrumental activities of daily living (such as housekeeping and preparing food), and robots assisting with enhanced activities of daily living (such as communication and engaging in hobbies) [10]. For example, "Care-O-bot" [12–14], a robot developed by Fraunhofer IPA, is able to fetch and carry objects, communicate with older people, and supply emergency support; "RIBA" [15, 16], a robot developed by RIKEN-TRI Collaboration Center, can help patient transfer; "uBot5" [17], another robot from the University of Massachusetts Amherst, is capable of achieving multiple postures for the purpose assisting elderly in compensating for impaired upper extremity function; "PerMMA (i.e., Personal Mobility and Manipulation Appliance)" [18–20], a research outcome from the Carnegie Mellon University and the University of Pittsburgh, can assist persons with severe physical disabilities; "PaPeRo" [21–23], another case developed by NEC, is used to communicate; and "Emiew" [24–26] developed by Hitachi, can interact with human beings; and "Hospi-R" [27, 28], an autonomous delivery robot developed by Matsushita, can even perform complex service tasks. The promising results of the above-mentioned AAL robotic systems indicated that robots could be used as effective ATS tools.

3 Background of Artificial Neural Networks

The original inspiration for ANN is the highly interconnected, massively parallel, distributed, adaptive neuron-and-synapse structure of the brain [29]. In general, the neurons are organized in the form of layers. Each neuron in a layer has weighted connections to neurons in other layers. The main working principle is the weights are adjusted adaptively according to the task under execution in order to improve the overall system performance. According to [30], there are three primary features of ANN, namely utilization of large amounts of sensory information, collective processing capability, and learning and adaptation capability. Broadly, ANN can be classified into two categories: single layer perceptron (SLP) and multi-layer perceptron (MLP).

3.1 Single Layer Perceptron

This network consists of two layers (see Fig.1): the input layer where data or signals are presented to the network, and the output layer which produces the output value of the network to a given input. That means, neurons grouped in layers with only connections between neurons in subsequent layer. Such a network is called a "single-layer", since there is only output layer that performs the computation. It is the basic form of a neural network used for the classification of a special type of

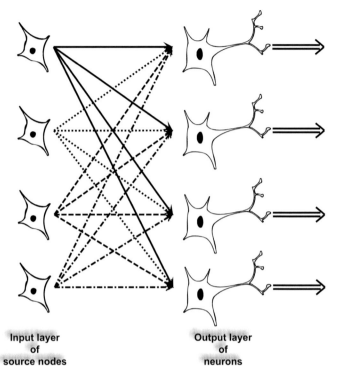

Fig. 1 Feedforward network with a single layer of neurons

patterns, which are linearly separable, i.e., patterns that lie on opposite sides of a hyperplane [31].

In its simplest form, it would consist of set of single inputs to a single neuron, which outputs a single output. A more useful form is illustrated in Fig. 2.

In Fig. 2, a signal x_j at the input of synapse j connected to neuron k is multiplied by the synaptic weight w_{kj}. It also adds the concept of an applied bias, denoted by b_k. It is used to increase or lower the net input of the activation function, depending on whether it is positive or negative, respectively.

In mathematical terms, we may describe the neuron k by writing the pair of equations as shown in Eqs. (1)–(3) [32]:

$$\mu_k = \sum_{j=1}^{m} w_{kj} \cdot x_j \tag{1}$$

$$y_k = \varphi(v_k) \tag{2}$$

$$v_k = \mu_k + b_k \tag{3}$$

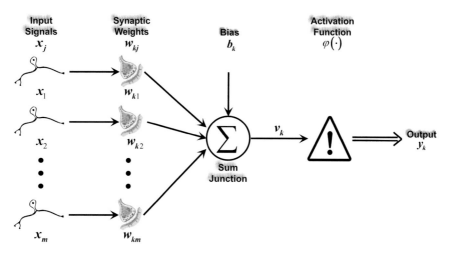

Fig. 2 Nonlinear model of a neuron, labeled k

where $x_1, x_2, \ldots, x_m$ are input signals; $w_{k1}, w_{k2}, \cdots, w_{km}$ are the respective synaptic weights of neuron k; μ_k is the linear combiner output to the input signals; b_k is the bias; $\varphi(\cdot)$ is the activation function which defines the output of a neuron in terms of the induced local field v; v_k is the activation potential or induced local field, and y_k is the output signal of the neuron.

In the simplest case of activation function (i.e., threshold function), the output is computed as Eq. (4) [32]:

$$\varphi(v_k) = \begin{cases} 1 & \text{if } v_k \geq 0 \\ 0 & \text{if } v_k < 0 \end{cases} \tag{4}$$

Correspondingly, the output of neuron k employing such a threshold function is expressed as Eq. (5) [32]:

$$y_k = \begin{cases} 1 & \text{if } v_k \geq 0 \\ 0 & \text{if } v_k < 0 \end{cases} \tag{5}$$

It is important to understand that the form of the activation function, once it is chosen, is will be used for all neurons in the network. In addition, the bias can be seen as an external parameter of neuron k. So we may formulate it as Eq. (6) [32]:

$$v_k = \sum_{j=0}^{m} w_{kj} \cdot x_j \tag{6}$$

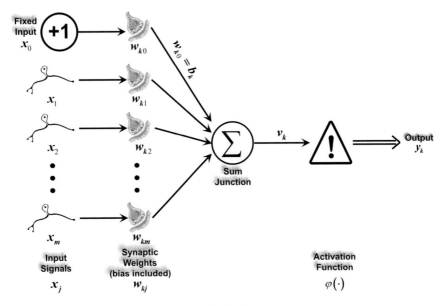

Fig. 3 Another nonlinear model for neuron, labeled k

In Fig. 3 a new synapse is added. Its input is shown in Eq. (7) [32]:

$$x_0 = +1 \qquad (7)$$

and its weight is expressed in the form of Eq. (8) [32]:

$$w_{k0} = b_k \qquad (8)$$

We may therefore reformulate the model of single neuron k as illustrated in Fig. 3.

3.2 Multi-layer Perceptron

In the literature, a feedforward neural network (FNN) with hidden layers is often referred to as multi-layer perceptron (MLP). The term "hidden" refers to all other layers with no direct connections from or to the outside. The function of hidden neuron is to intervene between the external input and the network output in some useful manner. According to [32], the basic features of MLPs are: (i) there are one or more layers that are hidden from both the input and output nodes; (ii) the network exhibits a high degree of connectivity; and (iii) the model of each neuron in the network has a nonlinear activation function which is differentiable. The performance of the multi-layer neural network is influenced greatly by the number of hidden layers and the number of nodes in a hidden layer. By adding one or more hidden layers, the network has the capability to extract higher-order statistics from its input. Overall, the output of MLPs are calculated through the following steps [32]:

First, the weighted sums of inputs are computed via Eq. (9) [32]:

$$v_j(n) = \sum_{i=0}^{m} w_{ji}(n) \cdot y_i(n) \tag{9}$$

where $v_j(n)$ is the induced local field produced at the input of the activation function associated with neuron j; m is the total number of inputs (excluding the bias) applied to neuron j; w_{ji} shows the connection weight from the ith node in the input layer to the jth node in the hidden layer; the synaptic weight w_{j0} (corresponding to the fixed input $y_0 = +1$) equals the bias b_j applied to neuron j. Hence, the function signal $y_j(n)$ appearing at the output of neuron j at iteration n is obtained via Eq. (10) [32]:

$$y_j(n) = \varphi_j\big(v_j(n)\big) \tag{10}$$

Then the induced local field for neuron k is expressed in Eq. (11) [32]:

$$v_k(n) = \sum_{i=0}^{m} w_{kj}(n) \cdot y_j(n) \tag{11}$$

where m denotes the total number of inputs (excluding the bias) applied to neuron k; the synaptic weight $w_{k0}(n)$ is equal to the bias $b_k(n)$ applied to neuron k; and the corresponding input is fixed at the value +1.

Next, if neuron j is in the first hidden layer of the network, then $m = m_0$ and the index i refers to the ith input terminal of the network, for which we can write Eq. (12) [32]:

$$y_i(n) = x_i(n) \tag{12}$$

where $x_i(n)$ is the ith element of the input vector (pattern).

On the other hand, if, neuron j is in the output layer of the network, then $m = m_L$ and the index j refers to the jth output terminal of the network, for which Eq. (13) can be obtained [32]:

$$y_j(n) = O_j(n) \tag{13}$$

where $O_j(n)$ is the jth element of the output vector of the MLP. This output is compared with the desired response $d_j(n)$, obtaining the error signal $e_j(n)$ for the jth output neuron.

Finally, the activation function commonly used in MLP is sigmoidal nonlinearity. It is defined via Eq. (14) [32]:

$$\varphi_j\big(v_j(n)\big) = \frac{1}{1 + \exp\big(-av_j(n)\big)}, \quad a > 0 \tag{14}$$

where $v_j(n)$ is the induced local field of neuron j, and a is an adjustable positive parameter.

For a neuron j located in the output layer, i.e., $y_j(n) = O_j(n)$, we may express the local gradient for neuron j in the form of Eq. (15) [32]:

$$\begin{aligned}\delta_j(n) &= e_j(n) \cdot \varphi'\big(v_j(n)\big) \\ &= a \cdot \big[d_j(n) - O_j(n)\big] \cdot O_j(n) \cdot \big[1 - O_j(n)\big], \quad \text{neuron } j \text{ is an output node}\end{aligned} \tag{15}$$

where $O_j(n)$ is the function signal at the output of neuron j, and $d_j(n)$ is the desired response for it.

On the other hand, for an arbitrary hidden neuron j, we may express the local gradient through Eq. (16) [32]:

$$\begin{aligned}\delta_j(n) &= \varphi'\big(v_j(n)\big) \cdot \sum_k \delta_k(n) \cdot w_{kj}(n) \\ &= a \cdot y_j(n) \cdot \big[1 - y_j(n)\big] \cdot \sum_k \delta_k(n) \cdot w_{kj}(n), \quad \text{neuron } j \text{ is hidden}\end{aligned} \tag{16}$$

3.3 Artificial Neural Network in Robot Control

Many problems in robotics are unknown, stochastic, and highly non-linear dynamics which offer significant challenges to traditional control methods. To cope with those problems, in recent years, artificial neural network (ANN) based control technique has become more and more popular. A neural network is a massive system of parallel distributed processing elements connected in a graph topology [32]. Usually, the networks are designed to direct a manipulator (e.g., motor) based on sensor data. According to the literature [32–34] the key factors of using ANN control scheme for robotics include: (i) no mathematical process model or rule-based knowledge required; (ii) highly adaptation; (iii) learning capability, such as supervised or unsupervised learning; (iv) fault tolerance; (v) massive parallelism; and (vi) generalization.

Until now, ANN based control has been applied in different robots control areas. For example, the authors of [34] presented a systematic design methodology to a motion adaptive control based on ANN. In [35], a ANN-based controller was developed for the tracking control of an *n* rigid-link robot manipulator. In addition, paper [36] employed an observer-based adaptive wavelet neural network tracking control scheme to tackle problems such as system uncertainties, multiple time-delayed state uncertainties, and external disturbances. More recently, a new ANN-based control platform called spiking neural network (SNN) has been used for mobile robot controllers. It used pulse codings to incorporate spatial-temporal information in communication and computation, like real neurons do. Interested readers please refer to [37–39] for more details.

4 Problem Statement

There are many extant versions of ANNs, despite their discrepancies, they all share one common feature that is learning which emphasize an ANN's capability of improving its performance based on the accumulated experience. Briefly, the learning capability of an artificial neuron is achieved by adjusting the weights in accordance to the chosen learning algorithm. Similarly to biological neurons, ANNs have been equipped with mechanisms to adapt themselves to a set of given inputs. Normally, there are two types of learning here: supervised and unsupervised. In the supervised learning, the ANN is provided with feedbacks from an external source (i.e., supervisor). In the latter case, however, an ANN adapts itself to inputs (aka learn) without any extra external feedbacks [40]. In general, the approach that offers learning to an ANN is called trainer. A trainer takes in charge of training NNs to achieve the highest performance for new sets of given inputs. In the supervised learning, a training approach first provides ANNs with a set of data called training data. The trainer then adjusts the structural parameters of the ANN in each training step for the purpose of improving the performance. Once the training phase is accomplished, the trainer is omitted and ANN is ready to use. The trainer if thus often regarded as the most important component of any ANNs.

Typically, there are two types of training approaches in the literature: deterministic versus stochastic. In the first sort, back propagation (BP) and gradient-based methods are often employed. In such techniques, the training phase results in the same performance if the training samples remain consistent. The trainers in this group are mostly mathematical optimization techniques that aim to find the maximum performance (minimum error). In contrast, the trainers in the stochastic camp utilize stochastic optimization techniques to fulfil the goal of maximizing performance of an ANN. The advantages of the deterministic trainers lie in their simplicity and converging speed. Deterministic training algorithm normally starts with a solution and guides it toward an optimum. Though the convergence speed is indeed very fast, quality of the obtained solution highly depends on the initial solution input. Also, there is a high probability of trapping in a local optima which is often referred to the sub-optimal solutions in a search space that might misleading us that the global optimum is obtained. The daunting issue here is the unknown number of runs that a trainer needs to be restarted with different initial solutions so as to increase the hope of finding global optimum. On the contrary, stochastic trainers start the training process with random solution(s) and evolve it (them). Randomness is the essential component of the stochastic trainers that apply to both initial solutions and method of solution's improvement during the training process. Though they are generally much slower than their deterministic counterparts, the advantage of stochastic methods is the high capability of local optima avoidance. This merit verifies the reasons of why stochastic training methods have been gaining much attention recently [40].

We can roughly divide stochastic training algorithms (in the context of training ANNs) into two groups: single-solution and multi-solution based algorithms. In the first sort, a trainer starts the training with a single randomly constructed ANN and evolves it repeatedly until the stopping criterion are satisfied. Examples of single-solution-based trainers are such as simulated annealing and hill climbing. In the second camp, a multiple-solution-based algorithm initiates the training process by a set of randomly created ANNs and evolves them with respect to each other until the termination conditions are met. The literature indicates that stochastic algorithms with multiple solutions often offer higher ability of escaping from local optima trap. Some of the most popular multi-solution trainers in the literature can be further grouped into two classes: conventional and innovatory metaheuristics [41]. The word heuristic has its origin in the old Greek word *heuriskein*, which means the art of discovering new strategies (rules) to solve problems. The suffix *meta,* also a Greek word, means "upper level methodology." Metaheuristic search approaches can be defined as upper level general methodologies (templates) that can be used as guiding strategies in designing underlying heuristics to solve specific optimization problems [42].

- Conventional metaheuristics: genetic algorithm (GA), particle swarm optimization (PSO), ant colony optimization (ACO), and differential evolution (DE)
- Innovatory metaheuristics: artificial bee colony (ABC), hybrid central force optimization and particle swarm optimization (CFO-PSO), social spider optimization (SSO) algorithm, chemical reaction Optimization (CRO), charged system search (CSS), invasive weed optimization (IWO), and teaching-learning based optimization (TLBO).

The reason as to why these optimization techniques have been employed as training algorithms is their outstanding performance in terms of approximating the global optimum.

To summarize, although many applications of ANNs (e.g., robot control) can be found in the literature, training ANNs is always a challenging task. As we can see from previous discussion, the weights and biases are responsible for defining the final output of MLPs from given inputs. Finding proper values for weights and biases in order to achieve a desirable relation between the inputs and outputs is the core of training MLPs. Bearing this in mind, the focal problem of this chapter is placed in the background of AAL with a focus of training MLPs for the purpose of controlling robot swarm.

5 Employed Methodology

In order to meet the chapter theme, we employ a newly developed algorithm called grey wolf optimizer (GWO) [43] which is inspired by the grey wolves hunting and searching behaviours. In [43], the authors classified the wolves into 4 groups, i.e., alphas, betas, deltas, and omegas. In addition, they assumed that among the groups, alpha (the fittest solution), beta, and delta have better knowledge about the potential location of prey [43]. Overall, the GWO can be seen as a two-stage method, i.e., encircling the prey during the hunting process using hyper-cubes framework and then employing an intensive local search mechanism for optimization. Like many other novel CI algorithms, GWO also includes a balance between exploitation/exploration. This offers the advantage of enhanced search ability while maintaining adequate exploitation capability.

5.1 Fundamentals of GWO

In the following, we describe the steps to be taken for obtaining an efficient implementation of GWO.

- Step 1: Generate the initial the grey wolf population, X_i, $i = 1, 2, \ldots, n$.
- Step 2: Initialize the algorithm parameters $(a, A, \text{ and } C)$ as follows [43]:

$$\begin{aligned} \vec{A} &= 2\vec{a} \cdot \vec{r}_1 - \vec{a} \\ \vec{C} &= 2 \cdot \vec{r}_2 \end{aligned} \tag{17}$$

where $\vec{A}$ and $\vec{C}$ are coefficient vectors, the components of $\vec{a}$ are linearly decreased from 2 to 0 over the course of iterations, and $\vec{r}_1$ and $\vec{r}_2$ are random vectors in $[0, 1]$.

- Step 3: Evaluating the fitness value, i.e., X_α, X_β, and X_δ.
- Step 4: Position correction-cooperation between current search agents by Eqs. (18) and (19), respectively [43]:

$$\begin{cases} \vec{D}_\alpha &= \left| \vec{C}_1 \cdot \vec{X}_\alpha - \vec{X} \right| \\ \vec{D}_\beta &= \left| \vec{C}_2 \cdot \vec{X}_\beta - \vec{X} \right| \\ \vec{D}_\delta &= \left| \vec{C}_3 \cdot \vec{X}_\delta - \vec{X} \right| \end{cases} \tag{18}$$

$$\begin{cases} \vec{X}_1 = \vec{X}_\alpha - \vec{A}_1 \cdot \left(\vec{D}_\alpha \right) \\ \vec{X}_2 = \vec{X}_\beta - \vec{A}_2 \cdot \left(\vec{D}_\beta \right) \\ \vec{X}_3 = \vec{X}_\delta - \vec{A}_3 \cdot \left(\vec{D}_\delta \right) \end{cases} \qquad (19)$$

where $\vec{D}$ is the distance of each candidate solution from the prey, $\vec{X}_\alpha, \vec{X}_\beta$, and $\vec{X}_\delta$ are the positions vector of the prey, t represents the current iteration, and $\vec{X}$ indicates the position vector of a grey wolf.

- Step 5: Updating the best location of the hunting wolves through Eq. (20) [43]:

$$\vec{X}(t+1) = \frac{\vec{X}_1 + \vec{X}_2 + \vec{X}_3}{3} \qquad (20)$$

- Step 6: Evaluating the stopping criteria. If yes, generate output; otherwise, go back to Step 2.

Although the GWO is designed in a very simple manner, i.e., only three main parameters need to be adjusted, each parameter has its own functionalities. For example, the objective for parameters a and A is to find a reasonable balance between two factors: first, a too narrow focus of the search process, which in the worst case may lead to stagnation; second, a too weak guidance of the search, which can cause excessive exploration, i.e., when $|A| < 1$, the wolves will attack towards the prey; otherwise, the wolves keep searching for prey. In addition, parameter C has two features: First, it provides random weights for prey in order to stochastically emphasize $(C > 1)$ or deemphasize $(C < 1)$ the effect of prey in defining the distance; Second, it represents the effect of obstacles to approaching prey in nature [43].

5.2 Training MLP via GWO

In general, the following common design questions [42, 44], namely, the representation, the objective, and the evaluation/cost function, are the key to a successful implementation of all metaheuristics. Loosely speaking, the representation refers how a computer keeps the candidate solutions and objects that it handles in the process of searching for new ones; the objective stands for the goal that one is planning to achieve; and the evaluation/cost function denotes a way of verifying the quality of the obtained solution to the problem. In summary, these three pillars form a strategic troika which means one has to give a careful consideration to each element.

The meanings of this statement, in the context of training ANN via meta-heuristics, are threefold: first, formulating the problem of training MLP in an acceptable way of utilizing a particular metaheuristic; second, setting up the training purpose, i.e., reducing the difference between the desirable outputs and the obtained outcomes; third, defining a suitable evaluation function that can guide the search of an employed metaheuristic. The following subsections elaborate how this could be done.

5.2.1 Solution Representation

In practice, the alternative candidate solutions are often encoded via representation for manipulating purpose, and thus it is often regarded as one of the fundamental design questions in the development of metaheuristics. For each tackled problem, the search space and its associated size are often determined by how a potential solution is represented and its corresponding interpretations. This statement implies that the efficiency and effectiveness of any implemented metaheuristic are often largely influenced by the chosen representation and the selected way of encoding, rather than by the problem itself. Picking out the right search pace is the first and foremost in implementing a metaheuristic. If an appropriate domain is not targeted at the very beginning of each search, one may fall into two embarrassing situations: either adding a large amount of unfeasible or repeated possible solutions to the shortlist, or excluding the potential right answer(s) from the selected search space.

Although the specific solution representation is often scenario dependent, some general rules can still be concluded as follows [42]:

- Comprehensiveness: A complete set of all possible solutions associated with the tackled problem have to be represented.
- Convexity: There is always a search path exists between any two solutions of the search space.
- Efficacy: The easy-to-manipulate (e.g., time and space complexities) degree of a representation must be high.

In the literature, there are many straightforward encoding methods can be used for dealing with some conventional families of optimization problems (see Fig. 4). In practice, these representation can be combined to form new type of representations for addressing new scenarios.

According to [40], the variables in vector form are suitable for GWO, therefore weights and biases of an MLP are described as Eq. (21) [40]:

$$\mathbf{V} = \{\mathbf{w}, \boldsymbol{\theta}\} = \{w_{1,1}, w_{1,2}, \dots, w_{n,n}, \quad \theta_1, \theta_2, \dots, \theta_h\} \tag{21}$$

where the number of the input nodes is denoted by n, $w_{i,j}$ represents the connection weight from the ith to the jth node, θ_j shows the threshold bias of the jth hidden node.

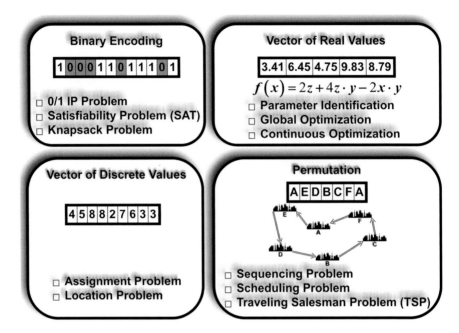

Fig. 4 Typical encoding examples

5.2.2 The Objective

Once the search space has been confirmed after the representation stage, one has to decide what the objective of the targeted problem at hand is. This is more like a task statement (in mathematical form) which concludes what need to be achieved. In the context of MLP training, the main purpose of training an MLP is to achieve the highest accurate degree in terms of classification, approximation, or predication for both training and testing samples. A commonly used measurement for quantifying such achievement is to calculate the mean square error (MSE), which means the difference between the desirable outputs (d_i) and the obtained outcomes (y_i), after a set of training samples being applied to an MLP. The objective of training an MLP can thus be interpreted simply as Eq. (22) [40]:

$$\min \sum (d_i - y_i)^2 \tag{22}$$

Mathematically, Eq. 23 [40] can be employed to calculate such difference.

$$MSE = \sum_{i=1}^{m} \left(d_i^k - y_i^k\right)^2 \tag{23}$$

where m denotes the number of outputs/outcomes, d_i^k and y_i^k refer to the desired output and actual outcome, respectively, of the ith input node when the kth training sample is applied.

5.2.3 The Evaluation Function

The evaluation function is generally not the same thing as the previously described objective. It is most commonly a mapping from the potential candidate solutions' space (under the selected representation scheme) to a set of numbers (e.g., the reals). In other words, the evaluation function $f(\cdot)$ associates with each candidate solution (belonging to the search space) a numeric value to indicate the quality or the fitness of the solution, e.g., $f : S \rightarrow \mathbb{R}$. The evaluation function gives one an opportunity to compare the usefulness of a present solution with its alternative counterparts. Sometimes, the evaluation function is ordinal since it only offers a ranking of all possible solutions; whereas it may also be numeric by offering not only the order of the solutions but also their corresponding quality degrees [42, 44].

The evaluation function is a crucial element in implementing a metaheuristic since it will guide the search towards "optimal" or "good-enough" solutions within the search space. In a word, regardless of whatever metaheuristic algorithm is employed, no feasible solutions can be obtained if the evaluation function is not properly defined. Nevertheless, in almost all real-world problems, the evaluation function does not come with the problem by itself. How should one go about making such choice? A good rule of thumb is a solution can be regarded as the best evaluation when it meets the objective thoroughly. In other words, it is unacceptable for an evaluation function to conclude that a solution (unable to meet the objective) is better than the other one (successfully meet the objective). But this is too elementary to design a suitable evaluation function.

Often, the objective implies a particular evaluation function. Based on our previous description, an MLP can be regarded as effective only if it is able to adapt itself to the whole training sample set. Accordingly, Eq. (24) below can be used to verify such effectiveness [40]:

$$\overline{\text{MSE}} = \sum_{k=1}^{s} \frac{\sum_{i=1}^{m} \left(d_i^k - y_i^k \right)^2}{s} \tag{24}$$

where s represents the number of training samples, and other parameters follow the same definition described in Eq. (7).

By now, the problem of training an MLP can be summarized and formulated as Eq. (25) below [40]:

$$\text{minimize} : f(\mathbf{V}) = \overline{\text{MSE}} \tag{25}$$

But, what if one cannot always derive valuable evaluation function from the objective? In these cases, one has to be clever enough to resort to some substitute evaluation functions that suit the needs of the problem at hand, the selected representation, and the operators that we implement to go from one solution to the next. The selection of such evaluation function is certainly out of the scope of this chapter, interested readers please refer to [44] for more details.

5.2.4 Training Progress

The generalized process of training an MLP via GWO is depicted in Fig. 5, from which we can see that GWO first takes the $\overline{\text{MSE}}$ from MLP, and then return MLP with the adjusted weights and biases. This process is performed recursively by GWO algorithm where the value of weights and biases are continuously rearranged so that the $\overline{\text{MSE}}$ can be largely minimized for all training data. An expanded version of this training process can be found in [40].

5.3 Concise Summary of Training Performance

As one of the most cited and test datasets, the Iris dataset [45], which comprises 4 attributes, 150 training/test samples, and 3 classes, is also employed to test the performance of training MLP through GWO. Due to the characteristics of Iris dataset, the MLP is constructed as a 4-9-3 structure and the problem thus has 75 variables in total. By testing the GWO-bolstered MLP trainer on the Iris dataset, we can obtain the following experimental results (see Fig. 6 for illustration): MSE = 0.229 ± 0.0032 and the classification rate equals to 92 %. The results verify the GWO's merits of superior local optimal avoidance capability, and outstanding accuracy achievability. Interested readers please refer to [40] for more evidence of GWO's advanced performance.

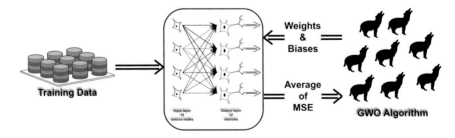

Fig. 5 Flowchart of GWO training MLP [40]

Fig. 6 Converging curve of
MLP trained by GWO

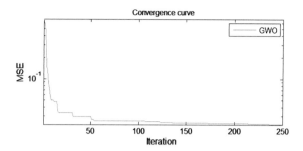

6 Experimental Environment Setup

6.1 Focal Scenario

Mobile robots can be roughly classified into the following groups: grounded mobile robots, nautical mobile robots, and aeronautical mobile robots. A complete AAL robot army can certainly involve all these members, say, grounded ones for interaction with elderly people, nautical ones for sewage cleaning, and aeronautical ones for house surroundings surveillance. Nevertheless, this seems too large for our experimental case to handle. So we decide to pay our attention only to the grounded mobile robots with a particular interest in human rescue task. The team members involved in our grouped mobile robots group are legged and wheeled robots. Typically, a heterogeneous robot team comprise distinct robots with different type of designs or functionalities that are most often complementary to each other for the purpose of accomplishing jobs more efficiently [46]. Unfortunately, the realization of a heterogeneous swarm of robots in real-world turns out to be an uphill work. The most daunting issue lies in that how the underlying control scheme could be designed and operated for dealing with the complex behavioural regulation and hardware arrangement [46]. Fortunately, by observing the behaviour of a swarm of social animals (e.g., ants, birds, herd, and fish) such as the features of self-organizing, decentralization, and emergence of collective actions, we can find inspiration of calibrating the desired robot team's control strategy [47]. For instance, the behaviour of ants [48] has inspired a number of researchers in pursuing collective robotics [49–53] design and analysis. An interesting initiative is "Swarmanoid" project [46, 54], which provides a modular framework for heterogeneous robot groups' control. Meanwhile, some researchers (e.g., [55]) also have explored heterogeneity at the hardware level (e.g., different sets of sensors or effectors). Bearing all this ongoing studies in mind, we make an attempt in this chapter to study the effectiveness of employing a novel control scheme, i.e., GWO algorithm trained MLP, in manipulating a heterogeneous swarm of ALL robots.

6.2 Mechatronic Design of Heterogeneous Robot Team

6.2.1 Design Blueprint

An important phase of building any robot swarm is its hardware design which is rather interactive since all components and/or parts will be used to assemble the final prototype. Since most existing work, at the hardware level, have been focusing on exploring the collective behaviour with homogeneous concept. Instead, in this project, we decided to exploit reconfigurability, and modularity using heterogeneous robots with decentralized control algorithms which are influenced by ants, bees colony, and insects behaviours [7]. One of the main characteristics of the designed heterogeneous robot swarm is that team members can perceive their surrounding environmental physical properties via various sensors, and can then executing auto-manipulation and fulfil self-localization through different actuators. The application of the developed robot warm is not only restricted to a research platform but also towards any potential deployment in real-world environments. The modular hardware architecture comprises independent sensory units, actuator modules, and communication kits, which help the swarm system to achieve the scalability and flexibility so that additional sensors and/or actuators could be added without bothering changing the overall architecture.

When designing and implementing the hardware platform for our targeted heterogeneous robot team, there are a lot of issues that need to be taken into account. Some of them are outlined as below:

- For wheel type of robots, a suitable diameter size for the shaft which houses the friction wheels.
- Select right type of wheels for avoiding slipping off while passing through special terrains, e.g., high inclines.
- Due to the rapid development of various techniques such as microcontroller, sensor units, and actuator kits, the platform's upgradability, modifiability, and compatibility need to be carefully considered.
- When integrating modularity and flexibility platform design concepts, one should not sacrifice the feature of user friendly.
- The reconfigurability of the platform should also be enhanced by fully embracing the software and middleware's upgrading.
- A sustainable power consumption plan, say, incorporating a high capacity battery which will lead to less power losses over the circuitry regulation.
- A wireless communication capability is a must for our robot team and thus a low cost but still effective information exchange plan should be investigated for both indoor and outdoor scenarios.
- The factors of size, shape, and weight of the potential platform also require an examination for the sake of robots' movability and manoeuvrability.
- The functionality, coordination, cooperation, and communicability between robots are also worth a scrutiny.

Bearing the abovementioned key points in mind, we set out to construct our heterogeneous robot team. By the time of compiling this chapter, we have preliminary completed two types of living assistant robots (wheel- and crocodile-robot) which are wholly assembled and ready to test for various applications such as simultaneously localization and mapping, avoiding obstacle, performing painting task, and executing rescue duty. The simplicity of the design allows an easy replication of each type of robot which can then form a swarm of robots. By utilizing local information and following some basic rules, these robot agents are able to sense, localize, and actuate for accomplishing complex work. For the rest of this section, the key modules of our robot (including both mechanical and electronic components) are briefed together with their working principles.

6.2.2 Input—Perception

Sensors are vital components of robotics system, since they provide information that allows us to monitor and to control the operation of these systems. Without the availability of sensory information, automated systems cannot operate. In our test environment, the functional movements of a limb involved for achieving a task are generally complex, therefore there is the need of combining different sensor modalities to improve the control process, such as proximity measurement (e.g., ultrasonic and infrared sensors), position measurement (e.g., encoders), temperature measurement (e.g., thermistors), vibration measurement (e.g., accelerometers), chemical measurement (e.g., gas sensors), GPS measurement, and video measurement (e.g., cameras) [56].

- Proximity measurement: Proximity sensors, used to determine the presence of nearby objects, were devised to extend the sensing range beyond that offered by direct-contact tactile sensors [57–59]. In our study, the main purpose of using these types of sensors is to detect obstacle and avoid collision. Based on specific properties used to initiate a switching action, they can be classified into several types: magnetic, inductive, ultrasonic, microwave, optical, and capacitive. Among these, ultrasonic sensors were found to be more accurate and have a much larger detection distance than other types of proximity sensors [59]. In addition, we employ infrared proximity sensors in motion detectors as well.

  - Ultrasonic sensors: Ultrasonic proximity sensors are available with an analog output voltage that is a function of the distance of the object away from the sensor or with two states of digital output that indicate object presence/absence within a defined zone. One feature of ultrasonic sensors is that they are not affected by the colour, transparency, or lighting conditions of the object being detected.
  - Infrared sensors: Infrared sensors consist of two parts: an infrared emitter and receiver. The emitter is actually an LED that emits light that's invisible to the human eye, and the receiver part of the switch collects the IR light that is reflected back. The working principle is that the obstacles cause more light

than usual to be reflected, which tells you that there is something in front of
the sensor.

- Position measurement: Position sensors are ones that that provide information
  about the change in the position of a rigid body [57–59]. These types of sensors
  can be classified as those that provide analog output (such as potentiometers and
  resolvers) and those that provide digital output (such as encoders).

  - Digital optical encoders: A digital optical encoder is a non-contact,
    optical-based device that converts motion into a sequence of digital pulses
    [60]. They have both linear and rotary configurations. The former are used to
    measure the linear position and velocity of a translating object, and the latter
    are used to measure the angular position and velocity for a rotating shaft.
    According to the literature [61], rotary encoders are the most widely used
    position sensors in robotic applications, since they provide acceptable res-
    olution with good noise immunity at low cost.

- Temperature measurement: Temperature is a basic quantity in process control
  systems, and there are several types of sensors available to measure temperature,
  such as thermistors, thermocouples, resistance temperature detector (RTD), and
  IC sensors.
- Vibration measurement: Vibratory motion commonly occurs in machinery and
  flexible structures. Measurement of vibration is important for machine health
  monitoring of motors, pumps, fans, gearboxes, machine tool spindles, blowers,
  and chillers. It is usually measured by either accelerometers or vibrometers.
- Chemical measurement: There are many sensors available for detecting chem-
  icals of various kinds, such as smoke, gas, moisture, etc.

  - Gas sensors: Gas sensors contain a small heating element and a catalytic
    detector to detect concentrations of gases. Using such a device involves
    supplying a voltage to the heating element and putting the sensors pins in a
    voltage divider arrangement with a fixed resistor to create a measureable
    output voltage.

- GPS measurement: GPS relies on a constellation of satellites. Each satellite
  contains a highly accurate clock that is synchronized with all the other satellites
  in the constellation. The satellites then broadcast this time signal.
- Cameras: Cameras are considered as complex environmental sensing systems
  that can be used for capturing and tracking objects locations, recognizing motion
  patterns, and so on.

6.2.3   Output—Action

In the context of mobile robots, mobility may refer to many actions such as the
maximum obstacle size that a mobile robot can get over, the steepest slope that a
mobile robot can go up, the largest amount of stairs that a mobile robot can climb,

the deepest swamp that a mobile robot can traverse, the highest distance that a mobile robot can jump, and the widest crevasse that a mobile robot can leap. But all these actions seem to be covered by a more generic concept, that is, mobility system which contains all necessary actuator components for constructing mobile robots. Since wheeled and legged mobile robots are the main focus of this case study, only key mobile system components needed by these robots are elaborated as below.

- Wheel size: The main objective in the control of wheeled robots is to remain balanced and avoid toppling [62]. In general, a bump which the mobile robots climb over, is one-third or less of the diameters of robot's wheels. As a result, to have the enough precession rate for balancing, design parameters of the wheel such as a diameter, a width, a mass, and spinning speed, which determine the size of the wheel that should be selected appropriately.
- Wheel structure: There are many structure alternatives for designing wheeled mobile robots, from more popular type with four-wheels to relatively unusual type with one-wheel. A robot with a single wheel has limited mobility, but enhances its obstacle-crossing ability, smoothness of motion and rolling efficiency [63]. Interested readers please refer to [64] for illustrative examples. For two wheels, there are two obvious layouts, i.e., bicycle-type, and inverted-pendulum type. The former is easier to control at low speeds, but it is not inherently stable; the latter is possible to achieve static stability by accurately placing the center of gravity on the wheel axle, but it always required for dynamic balancing. Three wheels are the minimum required for static stability. They come in many varieties, from very simple two-actuator differential drive, synchronous drive, to complicated holonomic omnidirectional type (e.g., an omnimobile robot with Swedish wheels, active caster wheels, and steerable wheels). Finally, the four wheeled layouts. The well-known one is the car-like structure, i.e., the front two wheels are synchronously steered to keep the same instantaneous center of rotation. A major advantage of this type is that it is stable during high-speed motion. However, it requires a slightly complicated steering mechanism.
- Leg geometry: Like wheeled robots, stability is a major concern in legged robots as well, since they tend to be tall and top heavy. As a result, the types of leg geometries are the key factors for the realization of proper walking capability.
- Leg actuator: One of the greatest problems in the realization of legged robots is the leg actuators which are responsible for the locomotion. In general, there are three major techniques for moving legs on a mobile robot, i.e., linear actuators, direct-drive rotary, and cable driven.

By now, we have our prototype robots (see Fig. 7) designed, built, and ready to be controlled.

6.2.4 Control—Inference

In general, mobile robot design problem can be classified into two directions: design of mechanical structure and development of an efficient control system.

Fig. 7 Prototype robot platform

Typically, robot control systems include open-loop and closed-loop. The architecture of an open-loop system is based on the current state and model of the system, whereas a control system that makes use of feedback is called a closed-loop system. Since the robots usually need to work in an unstructured environment, the closed-loop control system (i.e., with a means of responding to the problems) has been shown more powerful. In particular, if sufficient knowledge can be provided through an autonomous learning process, the control system can result in more autonomous and robust context. Nowadays, computational intelligence algorithms (e.g., artificial neural networks, genetic algorithm, and fuzzy logic) are well known to be computational tools to improve the performance of control techniques. Among others, artificial neural networks (ANNs) worth a mention due to their capability of learning, strong noise toleration, and generalizing. In this chapter, we employ supervised MLP (i.e., multi-layer perceptron) due to its popularity and simplicity (see Fig. 8 for illustration).

Fig. 8 Schematic representation of supervisied MLP

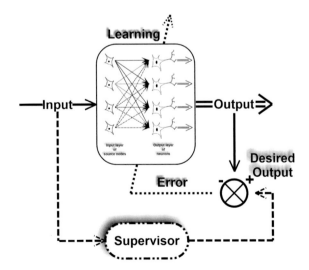

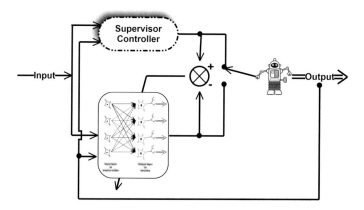

Fig. 9 Structure of neurocontrolled robot with supervised learning

The structure of supervise learning neuron robot control is shown in Fig. 9. As we can see from Fig. 9, the supervisor controller (i.e., GWO in the context of this study) trains the neurocontroller via different training patterns that can control the robot successfully. In detail, during the control period by the supervisor, each neuron is manipulated as a "black box" and all possible combination of neuron input signal and outputs/states of the robotic system are sampled, stored and analysed for the training of the neural network. After the training period, the neurocontroller takes the control actions, and the supervisor is deconnected from the system.

7   Limitations and Future Work

The limitations of the present work are threefold.

7.1   The Implementation of GWO Trained ANN into Heterogeneous Robot Control

In this chapter, we have successfully introduced the GWO trained MLP into our robot swarm control scenario. An immediate future work of the present study is to perform real-world testing on our robot team. The testing is envisioned as below: Initially only MLP controlled robots are employed and the time of completing the pulling task is recorded. Later on, GWO trained MLP control strategy should be introduced and the duration of task finishing is again recorded. For the purpose of evaluation, different time consumption and the achieved task accuracy can be compared between distinct robot deployments. Meanwhile interest readers can tailor their own arrangement to reach the best overall system performance.

7.2   Training ANN via Other Innovatory Nature Originated Algorithms

Recently, there are wide spread of nature originated metaheuristics across the literature. Nevertheless, according to our preliminary overview, the application of this algorithms in training MLP is limited. Apart from several attempts mentioned in the beginning of this chapter, there are still a large amount of approaches left untouched such as bat inspired algorithms, biogeography-based optimization algorithm, cat swarm optimization algorithm, cuckoo inspired algorithms, luminous insect inspired algorithms, fish inspired algorithms, frog inspired algorithms, fruit fly optimization algorithm, group search optimizer algorithm, music inspired algorithms, imperialist competitive algorithm, amoeboid organism algorithm, artificial search swarm algorithm, artificial tribe algorithm, bar systems algorithm, bean optimization algorithm, to name a few. Interested readers please refer to [41] for more details.

7.3   Comparing GWO Trained ANN with Other Versatile Versions of Nature Originated Algorithms in Robot Control

The third limitation of this study is the lack of the comparison between GWO trained ANN and the other nature originated algorithms in terms of robot control. At the end of this study, we briefly overview the corresponding literature with a hope of directing future research in this regard.

7.3.1   Bacteria Foraging Optimization in Robot Control

Bacterial foraging optimization (BFO) algorithm was originally proposed in Passino [65] where the foraging strategy of *E. Coli* bacteria has been simulated. Typically, the BFO consists of four main mechanisms: chemotaxis, swarming, reproduction, and elimination-dispersal event. Based on the basic BFO, an improved version was proposed in [66] to deal with robot navigation problem. Two simulation scenarios, namely, (i) fixed obstacle and target; and (ii) randomly moving obstacles and fixed target, were considered by the authors of [66]. 10 standard deviation values of path lengths were assigned to each testing scenarios where the obtained results are classified into three groups, i.e., best, worst, and average. By comparison the experiment results, both BFO algorithms (improved and basic version) outperform the traditional particle swarm optimization algorithm. At the end of their study, the authors claimed that BFO is very powerful in fulfilling the robot path navigation task.

7.3.2 Bees Algorithm in Robot Control

Inspired by natural foraging behavior of honey bees, the bees algorithm (BA) was proposed by the authors of [67]. Mobile robots for public service have always been a great research interest both for academic and industry. During the past few years, the study of robot swarm and its related topic has become a hot area. In [68], the authors utilized BA to study a group of reconfigurable mobile robots which are designed to provide daily service in hospital environments for different kinds of tasks such as guidance, cleaning, delivery, and monitoring. The fulfilment of each job requires an associated functional module that can be installed onto various robot platforms via a standard connection interface. Since the classic BA focuses mainly on single-objective functional optimization problems, a variant called binary BA (BBA for short) was proposed by the authors of [68] to deal with the multi-objective multi-constraint combinatorial optimization task. In BBA, a bee is describe as two binary matrixes **MR** and **RH**, standing for how to assign the M tasks to the R robots and the R robots to the H homes, respectively. The size of **MR** is $M \times R$ in which its R columns represent the R robots, while the M missions is represented by the M rows. The authors evaluated the BBA with an example problem (20 missions, 8 robots, and 4 homes) with a size of $8^{20} \times 4^8 = 2^{76}$ combinations. At first, 12 stochastic solutions are obtained by scout bees through global search in which six elite bees survive after the non-dominated selection. The final experiments demonstrated that BBA is a suitable candidate tool in treating workload balancing issue among a team of swarm robots.

7.3.3 Glowworm Swarm Optimization in Robot Control

The glowworm swarm optimization (GSO) algorithm was proposed by the authors of [69]. This algorithm gets its inspiration from the behavior from glowworms and it shares some common points with other population-based algorithms such as ant colony optimization and particle swarm optimization. The agents in GSO are a group of glowworms that carry a luminescent quantity call luciferin. Normally, GSO starts by randomly placing n glowworms in the search space so that they are well distributed. Initially, all glowworms carry an equal quantity of luciferin l_0. Typically, each iteration of GSO algorithm consists of three rules, namely, luciferin-update rule, movement rule, and transition rule.

In [70], the authors built a team of four wheeled-mobile robot named Kinbots, and employ GSO algorithm to fulfill the collective robot control. In the work, they equipped their Kinbots with infrared sensor-based interaction modules which can offer a hardware capability to perform luciferin emission or detection and a behavior of leader-following. In a subsequent study [71], the same authors conducted a preliminary experiment to demonstrate the performance of GSO in robot swarm control. A set of three glowworms (i.e., Kinbots) A, B, and C are initially located at the corners of an equilateral triangle (side is 50 cm). Kinbots A and B

remained stationary during the study, and emitted a luciferin value of 128 and 60, respectively, while a luciferin value of 40 was glowed by the Kinbot C. At each iteration, the sweep platform performs three tasks: first, homes by turning clock-wise until it make a proper angle with the heading direction of the Kinbots; second, does a 180° scanning to acquire intensity samples and localize the neighbors; and third, aligns along the line-of-sight of each neighbor to receive the luciferin value emitted by itself. The sensing phase of the first cycle is completed at 10 s. For simplicity, the authors of [71] introduced a maximum-neighbor selection rule that is a Kinbot chooses to move toward a neighbor which emits maximum luciferin. The simulation results demonstrated the suitability of applying GSO in dealing with the problem of multiple source localization encountered in the domain of robot control.

7.3.4 Electromagnetism-like Mechanism in Robot Control

Similar to that in the elementary electromagnetism, the authors of [72] regarded teach sample point as a charged particle that is released to a space. In the proposed electromagnetism-like mechanism (EM) approach, the objective function value is associated with the charge of each point which determines the magnitude of attraction/repulsion of the point over the sample population. In other words, the higher the magnitude of attraction, the better the objective function value. Once we get the value of these charges, we can use them to look for a direction, usually obtained through the evaluation of a combination force that exerts on the point via other points, where each point can move toward in the subsequent iterations. Like the electromagnetic forces, by adding vectorially the forces from each of the other points, we can obtain the required force. Typically, the EM algorithm consists of four phases, namely, initialization, local search, total force calculation, and the movement.

In order to deal with the path tracking problem, in [73], the authors employed EM to minimize the mobile robot controller's cost function in real time manner. The authors made the comparisons between two algorithm, namely, EM and the reference algorithm. From the simulation results, it is observed that the linear and angular velocities of the target mobile robot optimized by EM method have a faster convergence speed. The study of [73] successfully demonstrated that the EM approach present a good performance in minimizing the cost function. The major advantage of employing EM algorithm was, concluded at the end of [73], the ability to provide an effective and simple predictive control strategy.

7.3.5 Intelligent Water Drops Algorithm in Robot Control

The inspiration of intelligent water drops (IWD) algorithm comes from the water drops that flow into rivers, lakes, and seas. The core concept is that gravitational form of the earth drags the water drops in a river to flow towards their final destination. The author of [74] invented this algorithm in 2007. With the recent technological advancement, the development of unmanned vehicular systems, in particular the unmanned combat aerial vehicle (UCAV), have been proved to be beneficial in both military and civilian applications. Nevertheless, the complete benefits of such unmanned systems can only be fulfilled and utilized when their operations could achieve an autonomous level. One of the key requirements for realizing such autonomy is the ability of detecting internal and external changes, and reacting to them in a safe and efficient manner, especially without the intervention from their human operators. Under such circumstance, the trajectory planning becomes a nontrivial task. Typically, the goal of trajectory planning is to generate a space path between an initial location and the desired destination that has an optimal or near-optimal performance under different constraint conditions. In [75], the authors made an attempt to solve this imperative task by utilizing IWD algorithm. Since the generated UCAV optimal trajectory using the proposed IWD algorithm is normally difficult to be implemented in real flying environment due to the potential turning points on the optimized trajectory, the authors of [75] further adopted a class of dynamically feasible trajectory smooth strategy named k-trajectory. Finally, a series of case studies were conducted in their study under complicated combating environments. From the experimental results, it can be observed that the proposed IWD algorithm can find a feasible and optimal trajectory for the single UCAV. Meanwhile, the adopted k-trajectory approach is also every effective in smoothing the UCAV trajectory with a small computational load and real-time simulation possibility.

7.3.6 Gravitational Search Algorithm in Robot Control

Gravitational search algorithm (GSA) was originally proposed by Rashedi et al. [76]. In GSA, all the individuals can be mimicked as objects with masses. Based on the Newton's law of universal gravitation, the objects attract each other by the gravity force, and the force makes all of them move towards the ones with heavier masses. In addition, each mass of GSA has four characteristics: position, inertial mass, active gravitational mass, and passive gravitational mass. The first one corresponds to a solution of the problem, while the other three are determined by fitness function. More detailed discussions regarding GSA can be found in [41, 76].

Although legged robots are slower and more complicated, they have many advantages under certain conditions, such as better adaptability to irregular terrain conditions, and better climbing and obstacle overcrossing capability, which enable them to be more flexible than other types of locomotive mechanisms and can thus be deployed in multitude of dynamically changing situations. While building walking robot, stability is a key factor that needs to be considered since it is fundamental to the overall performance of terrestrial locomotion. In [77], the author made an attempt to use the characteristics of genetic and GSA to generate gait for a hexapod walking robot. The experimental results demonstrated that, in general, the increase of fitness of transformed gaits can be achieved in comparison with the fitness of the initial gait population. At the end of the study, the author of [77] suggested that supplementary mechanisms can be added for compensating the deviation of the robot path from pre-set trajectory.

8 Conclusion

The research of AAL robot has for sure not yet reached its frontiers and there is still a lot of work could be done to narrow the gap between academia's know-how and practitioners' requirements. In this work, MLP, an algorithm that is planned to be used for heterogeneous AAL robot team control, was trained via a new proposed GWO algorithm. In order to get desired results, the training problem was first formulated to fit the needs of GWO algorithm. The optimal values for weights and biases of MLP are then determined by GWO. The capability of performing high level of exploration and exploitation proved GWO's usefulness of training MLP. Another contribution made by this study lies in that it also provided a picture about some newly developed nature originated metaheuristic algorithms and how are they being applied to smart robot control area. Though this work is not completed without certain limitations, the study itself is exploratory in nature. It is believed that this chapter can, through the scattered literature, open a new window to other scholars who share the similar research interests.

References

1. Anonymous.: Ambient Assisted Living Roadmap. European Ambient Assisted Living Innovation Alliance. IOS Press, Amsterdam, The Netherland (2010)
2. Borja, R., de la Pinta, J.R., Álvarez, A., Maestre, J.M.: Integration of service robots in the smart home by means of UPnP: a surveillance robot case study. Robot. Auton. Syst. **61**, 153–160 (2013)
3. Schauer, D., Hein, A., Lueth, T.C.: RoboPoint—an autoclavable interactive miniature robot for surgery and interventional radiology. Int. Congr. Ser. **1256**, 555–560 (2003)
4. Pisla, D., Gherman, B., Vaida, C., Suciu, M., Plitea, N.: An active hybrid parallel robot for minimally invasive surgery. Robot. Comput. Integr. Manuf. **29**, 203–221 (2013)
5. Yu, H., Huang, S., Chen, G., Thakor, N.: Control design of a novel compliant actuator for rehabilitation robots. Mechatronics **23**, 1072–1083 (2013)
6. Meng, W., Liu, Q., Zhou, Z., Ai, Q., Sheng, B., Xie, S.S.: Recent development of mechanisms and control strategies for robot-assisted lower limb rehabilitation. Mechatronics (in press)
7. Banks, M.R., Willoughby, L.M., Banks, W.A.: Animal-assisted therapy and loneliness in nursing homes: use of robotic versus living dogs. J. Am. Med. Directors Assoc. **9**, 173–177 (2008)
8. Tzafestas, S.G.: Introduction to Mobile Robot Control. Elsevier Inc., London (2014), ISBN 978-0-12-417049-0
9. Wu, Y.-H., Wrobel, J., Cristancho-Lacroix, V., Kamali, L., Chetouani, M., Duhaut, D., et al.: Designing an assistive robot for older adults: the ROBADOM project. IRBM **34**, 119–123 (2013)

10. Rashidi, P., Mihailidis, A.: A survey on ambient-assisted living tools for older adults. IEEE J. Biomed. Health Inform. **17**, 579–590 (2013)
11. Feil-Seifer, D., Matarić, M.J.: Defining socially assistive robotics. In: Proceedings of the 2005 IEEE 9th International Conference on Rehabilitation Robotics, June 28–July 1, 2005. Chicago, IL, USA (2005)
12. Graf, B.: (2014). Care-O-bot. http://www.care-o-bot.de/en/care-o-bot-3.html. Accessed on 30 July 2015
13. Graf, B., Hans, M., Schraft, R.D.: Care-O-bot II: development of a next generation robotic home assistant. Auton. Robots **16**, 193–205 (2004)
14. Graf, B., Parlitz, C., Hägele, M.: Robotic home assistant Care-O-bot 3 product vision and innovation platform. In: Jacko JA (ed.) Human-Computer Interaction, Part II, (HCII 2009), LNCS 5611, pp. 312–320. Springer, Berlin (2009)
15. RIKEN-TRI Collaboration Center.: RIBA. http://rtc.nagoya.riken.jp/RIBA/index-e.html. Accessed on 30 July 2015
16. Mukai, T., Hirano, S., Nakashima, H., Kato, Y., Sakaida, Y., Guo, S., et al.: Development of a nursing-care assistant robot RIBA that can lift a human in its arms. In: Presented at the The 2010 IEEE/RSJ International Conference on Intelligent Robots and Systems, October 18–22, 2010. Taipei, Taiwan (2010)
17. Kuindersma, S.R., Hannigan, E., Ruiken, D., Grupen, R.A.: Dexterous mobility with the uBot-5 mobile manipulator. In: Presented at the International Conference on Advanced Robotics (ICAR), June 2009, pp. 1–7 (2009)
18. Xu, J., Grindle, G.G., Salatin, B., Vazquez, J.J., Wang, H., Ding, D., et al.: Enhanced bimanual manipulation assistance with the personal mobility and manipulation appliance (PerMMA). In: Presented at the The 2010 IEEE/RSJ International Conference on Intelligent Robots and Systems, October 18–22, 2010. Taipei, Taiwan (2010)
19. Wang, H., Grindle, G.G., Candiotti, J., Chung, C., Shino, M., Houston, E., et al.: The personal mobility and manipulation appliance (PerMMA): a robotic wheelchair with advanced mobility and manipulation. In: Presented at the The 34th Annual International Conference of the IEEE EMBS, San Diego, California USA, 28 Aug–1 Sept 2012
20. Cooper, R.A., Grindle, G.G., Vazquez, J.J., Xu, J., Wang, H., Candiotti, J., et al.: Personal mobility and manipulation appliance-design, development, and initial testing. Procddings IEEE **100**, 2505–2511 (2012)
21. Sato, M., Sugiyama, A., Ohnaka, S.: Auditory system in a personal robot, PaPeRo. In: 2006 Digest of technical Papers International Conference on Consumer Electronics (ICCE 06), pp. 19–20. 7–11 Jan 2006
22. Sato, M., Iwasawa, T., Sugiyama, A., Nishizawa, T., Takano, Y.: A single-chip speech dialogue module and its evaluation on a personal robot, PaPeRo-mini. In: Presented at the IEEE International Conference on Acoustics, Speech and Signal Processing (ICASSP), pp. 3697–3700, 19–24 April. Taipei, Taiwan (2009)
23. Fujiwara, N., Hagiwara, Y., Choi, Y.: Development of a learning support system with PaPeRo. In: Presented at the The 12th International Conference on Control, Automation and Systems, pp. 1912–1915, 17–21 October. Jeju Island, Korea (2012)
24. Hosoda, Y., Yamamoto, K., Ichinose, R., Egawa, S., Tamamoto, J.: Collision-avoidance algorithm for human-symbiotic robot. In: Presented at the International Conference on Control, Automation and Systems 2010, pp. 557–561, 27–30 October. Gyeonggi-do, Korea (2010)
25. Hosoda, Y., Egawa, S., Tamamoto, J., Yamamoto, K., Nakamura, R., Togami, M.: Basic design of human-symbiotic robot EMIEW. In: Presented at the Proceedings of the 2006 IEEE/RSJ International Conference on Intelligent Robots and Systems, pp. 5079–5084, 9–15 October. Beijing, China (2006)
26. HITACHI.: Robotics: EMIEW 2. http://www.hitachi.com/rd/portal/research/robotics/emiew2_01.html (2014). Accessed on 30 July 2015
27. Falconer, J.: HOSPI-R drug delivery robot frees nurses to do more important work. http://www.gizmag.com/panasonic-hospi-r-delivery-robot/29565/ (2013). Accessed on 30 July 2015

28. Murai, R., Sakai, T., Kawano, H., Matsukawa, Y.: A novel visible light communication system for enhanced control of autonomous delivery robots in a hospital. In: Presented at the IEEE/SICE International Symposium on System Integration (SII), pp. 510–516, 16–18 December. Kyushu University, Fukuoka, Japan (2012)
29. Russell, S.J., Norvig, P.: Artificial Intelligence: A Modern Approach, 3rd edn. Pearson Education, Inc., Upper Saddle River (2010), ISBN 978-0-13-604259-4
30. Haykin, S.: Neural Networks: A Comprehensive Foundation, 2nd edn. Pearson Education, Inc., Delhi, India, (1999), ISBN 8I-7808-300-0
31. Rosenblatt, F.: Principles of Neurodynamics. Spartan Books, Washington, DC (1962)
32. Haykin, S.: Neural Networks and Learning Machines, 3rd edn. Pearson Education, Inc., Upper Saddle River (2009), ISBN 978-0-13-147139-9
33. Erdem, H.: Application of neuro-fuzzy controller for sumo robot control. Expert Syst. Appl. **38**, 9752–9760 (2011)
34. Patiño, H.D., Carelli, R., Kuchen, B.R.: Neural networks for advanced control of robot manipulators. IEEE Trans. Neural Networks **13**, 343–354 (2002)
35. Wai, R.-J.: Tracking control basedon neural network strategy for robot manipulator. Neurocomputing **51**, 425–445 (2003)
36. Yu, W.-S., Weng, C.-C.: An observer-based adaptive neural network tracking controlof robotic systems. Appl. Soft Comput. **13**, 4645–4658 (2013)
37. Oniz, Y., Kaynak, O.: Control of a direct drive robot using fuzzy spiking neural networks with variable structure systems-based learning algorithm. Neurocomputing **149**, 690–699 (2015)
38. Wang, X., Hou, Z.-G., Zou, A., Tan, M., Cheng, L.: A behavior controller based on spiking neural networks for mobile robots. Neurocomputing. **71**, 655–666 (2008)
39. Wang, X., Hou, Z.-G., Lv, F., Tan, M., Wang, Y.: Mobile robots'modular navigation controller using spiking neural networks. Neurocomputing **134**, 230–238 (2014)
40. Mirjalili, S.: How effective is the grey wolf optimizer in training multi-layer perceptrons. Appl. Intell. **43**, 150–161 (2015)
41. Xing, B., Gao, W.-J.: Innovative Computational Intelligence: A Rough Guide to 134 Clever Algorithms. Springer International Publishing Switzerland, Cham (2014), ISBN 978-3-319-03403-4
42. Talbi, E.-G.: Metaheuristics: From Design to Implementation. Wiley, Hoboken (2009), ISBN 978-0-470-27858-1
43. Mirjalili, S., Mirjalili, S.M., Lewis, A.: Grey wolf optimizer. Adv. Eng. Softw. **69**, 46–61 (2014)
44. Michalewicz, Z., Fogel, D.B.: How to Solve it: Modern Heuristics, 2nd edn. Springer, Berlin (2004), ISBN 3-540-22494-7
45. Kuncheva, L.: Combining Pattern Classifiers: Methods and Algorithms, 2nd edn. Wiley, Hoboken (2014), ISBN 978-1-118-31523-1
46. Dorigo, M., Floreano, D., Gambardella, L. M.F., Mondada, F., Nolfi, S., Baaboura, T., et al: Swarmanoid: a novel concept for the study of heterogeneous robotic swarms. IEEE Robot. Autom. **20**, 60–71 (2013)
47. Dorigo, M., Stützle, T.: Ant Colony Optimization. The MIT Press, Cambridge (2004), ISBN 0-262-04219-3
48. Deneubourg, J.L., Goss, S., Franks, N., Sendova-Franks, A., Detrain, C., Chretien, L.: The dynamics of collective sorting robot-like ants and ant-like robots. In: Presented at the Proceedings of 1st Conference on Simulation of Adaptive Behavior (1991)
49. Kube, C.R., Bonabeau, E.: Cooperative transport by ants and robots. Robot. Auton. Syst. **30**, 85–101 (2000)
50. Holland, O., Melhuish, C.: Stigmergy, self-organization, and sorting in collective robotics. Artif. Life **5**, 173–202 (1999)
51. Caro, G.D.: A society of ant-like agents for adaptive routing in networks. Unpublished Master Thesis, Universite Libre de Bruxelles, Brussels, Belgium (2002)
52. Dorigo, M.: Swarms of self-assembling robots. In: Weyns D., Brueckner S.A., Demazeau Y. (eds.) EEMMAS 2007, LNAI 5049, pp. 1–2. Springer, Berlin (2008)

53. Dorigo, M., Tuci, E., Trianni, V., Groß, R., Nouyan, S., Ampatzis, C., et al.: SWARM-BOT: design and implementation of colonies of self-assembling robots. In: Yen G.Y., Fogel D.B. (eds.) Computational Intelligence: Principles and Practice, pp. 103–135. IEEE Computational Intelligence Society, New York (2006)
54. Ferrante, E.: A control architecture for a heterogeneous swarm of robots: the design of a modular behavior-based architecture. Doctor of Philosophy, Universite Libre de Bruxelles (2009)
55. Brunete, A., Hernando, M., Gambao, E., Torres, J.E.: A behaviour-based control architecture for heterogeneous modular, multi-configurable, chained micro-robots. Robot. Auton. Syst. **60**, 1607–1624 (2012)
56. Siegwart, R., Nourbakhsh, I.R.: Introduction to Autonomous Mobile Robots. The MIT Press, Cambridge (2004), ISBN 0-262-19502-X
57. Regtien, P.P.L.: Sensors for Mechatronics. Elsevier Inc., London (2012), ISBN 978-0-12-391497-2
58. Scherz, P., Monk, S.: Practical Electronics for Inventors. McGraw-Hill, New York (2013), ISBN 978-0-07-177134-4
59. Jouaneh, M.: Fundamentals of mechatronics. Cengage Learning, Stamford (2013), ISBN 978-1-111-56901-3
60. Sinclair, I.R., Dunton, J.: Practical Electronics Handbook, 6th edn. Newnes, Elsevier Ltd., Oxford (2007), ISBN 978-0-75-068071-4
61. Kurfess, T.R. (ed.): Robotics and Automation Handbook. CRC Press LLC, Danvers (2005), ISBN 0-8493-1804-1
62. Chan, R.P.M., Stol, K.A., Halkyard, C.R.: Review of modelling and control of two-wheeled robots. Annu. Rev. Control **37**, 89–103 (2013)
63. Xu, Y., Ou, Y.: Control of Single Wheel Robots. Springer, Berlin (2005), ISBN 978-3-540-28184-9
64. Park, J.H., Jung, S.: Development and control of a single-wheel robot: practical mechatronics approach. Mechatronics **23**, 594–606 (2013)
65. Passino, K.M.: Biomimicry of bacterial foraging for distributed optimization and control. IEEE Control Syst. Manage. **22**, 52–67 (2002)
66. Hossain, M.A., Ferdous, I.: Autonomous robot path planning in dynamic environment using a new optimization technique inspired by bacterial foraging technique. Robot. Auton. Syst. **64**, 137–141 (2015) (in press)
67. Pham, D.T., Ghanbarzadeh, A., Koç, E., Otri, S., Rahim, S., Zaidi, M.: The bees algorithm—a novel tool for complex optimisation problems. In: Second International Virtual Conference on Intelligent production machines and systems (IPROMS), pp. 454–459 (2006)
68. Xu, S., Ji, Z., Pham, D.T., Yu, F.: Bio-inspired binary bees algorithm for a two-level distribution optimisation problem. J. Bionic Eng. **7**, 161–167 (2010)
69. Krishnanand, K.N., Ghose, D.: Glowworm swarm optimization for simultaneous capture of multiple local optima of multimodal functions. Swarm Intell. **3**, 87–124 (2009)
70. Krishnanand, K.N., Ghose, D.: Detection of multiple source locations using a glowworm metaphor with applications to collective robotics. In: IEEE Swarm Intelligence Symposium (SIS), pp. 84–91 (2005)
71. Krishnan, K.N., Amruth, P., Guruprasad, M.H., Bidargaddi, S.V., Ghose, D.: Glowworm-inspired robot swarm for simultaneous taxis towards multiple radiation sources. In: IEEE International Conference on Robotics and Automation (ICRA), pp. 958–963. May, Orlando, Florida, USA (2006)
72. Birbil, Şİ., Fang, S.-C.: An electromagnetism-like mechanism for global optimization. J. Global Optim. **25**, 263–282 (2003)
73. Wang, Y., Yang, Y., Yuan, X., Yin, F., Wei, S.: A model predictive control strategy for path-tracking of autonomous mobile robot using electromagnetism-like mechanism. In: International Conference on Electrical and Control Engineering (ICECE), pp. 96–100 (2010)
74. Shah-Hosseini, H.: Problem solving by intelligent water drops. In: IEEE Congress on Evolutionary Computation (CEC), pp. 3226–3231, 25–28 September (2007)

75. Duan, H., Liu, S., Wu, J.: Novel intelligent water drops optimization approach to single UCAV smooth trajectory planning. Aerosp. Sci. Technol. **13**, 442–449 (2009)
76. Rashedi, E., Nezamabadi-pour, H., Saryazdi, S.: GSA: a gravitational search algorithm. Inf. Sci. **179**, 2232–2248 (2009)
77. Seljanko, F.: Hexapod walking robot gait generation using genetic-gravitational hybrid algorithm. In: 15th International Conference on Advanced Robotics, pp. 253–258, 20–23 June. Tallinn University of Technology, Tallinn, Estonia (2011)

Part II
Control Tuning and Adaptive Control Systems

Evolutionary Modeling of Industrial Plants and Design of PID Controllers

Case Studies and Practical Applications

Monica Patrascu and Andreea Ion

Abstract This chapter brings forth the practical aspect of using genetic algorithms (GAs) in aiding PID (Proportional-Integral-Derivative) control design for real world industrial processes. Plants such water tanks, heaters, fans and motors are usually hard to tune on-site, especially after prolonged use of the equipment when degradation of performances is inevitable, while plants like seismic dampers have inherent nonlinear behaviors that make formal controller design difficult at best. Therefore, this chapter introduces a series of practical steps that can be taken by control engineers in order to (re)design viable PID controllers for their plants. This chapter describes how genetic algorithms can be applied to problems in control systems and model identification. Considering the plant inputs and outputs that can be observed during functioning, we offer a quick method for identifying model parameters, which can be used later by the genetic algorithm to find a suitable controller. While, in formal control theory, a raw estimation of the model parameters can significantly reduce the performance of a real-world system, the genetic algorithm method can find suitable controllers quickly and efficiently, offering access even to performance criteria that is hard to quantify in classical design procedure, such as integral indexes. The applicability of the GAs in real world problems is outlined through case studies that take into account the particularities of each system, from first to second order responses, the absence or presence of time delay, nonlinearities, constraints and controller performance. The steps performed in the case studies show how GAs have made the jump from their origins to a practicing engineer's toolbox (GAOT-ECM in this case, a Genetic Algorithm Optimization Toolbox Extension for Control and Modeling). Moreover, a comprehensive analysis is performed, that takes into account both the various performance criteria, and the tuning parameters of the genetic algorithm, over the obtained models and controllers. The influence of the GA parameters is discussed in

M. Patrascu (✉) · A. Ion
Department of Automatic Control and Systems Engineering, University Politehnica
of Bucharest, Bucharest, Romania
e-mail: monica.patrascu@acse.pub.ro

A. Ion
e-mail: andreea.ion@acse.pub.ro

© Springer International Publishing Switzerland 2016
H.E. Ponce Espinosa (ed.), *Nature-Inspired Computing for Control Systems*,
Studies in Systems, Decision and Control 40, DOI 10.1007/978-3-319-26230-7_4

order to help practitioners choose the best suited GA configuration for their particular problem. In all, this chapter offers a comprehensive step-by-step application of genetic algorithms in industrial setting, from plant modeling to controller design.

Keywords Genetic algorithms · Plant modeling · Control systems · PID control

1 Introduction

From the beginning of the computer age, scientists have been oriented toward developing intelligent techniques to outline the adaptive capabilities of computers and software entities. Over the years, these research endeavors have materialized into a variety of methods like expert systems, neural networks, fuzzy techniques, intelligent agents, evolutionary systems. In the field of computer science, evolutionary algorithms (EAs) are a subfield of artificial intelligence that is defined by algorithms that are generally based on Darwin's "survival of the fittest" principle. Nowadays, a few categories of this field are: evolution strategies (ESs), genetic programming (GP), and genetic algorithms (GAs). From these classes, GAs and GP are particularly suitable in problems in which conventional optimizers are inefficient or inappropriate.

For this chapter, the considered EA class is genetic algorithms (GAs) [1], that are computer-based search methods patterned after the genetic evolution mechanisms of biological organisms that have adapted in highly competitive environments for millions of years. In technology and science, GAs have been used as adaptive algorithms and computational models of natural evolution systems [2]. GAs have been successfully applied to problems in various areas of study, and their popularity continues to increase because of their effectiveness, applicability, and ease of use. It has been shown that GAs are efficient for problems in control systems design [3] for both feedback [4, 5] and feedforward controllers [6], model selection [7, 8], and estimation of model parameters [9, 10]. GAs are used for a variety of problems, from experimental video identification [11], to instrument recognition [12], either stand-alone or in combinations with other intelligent methods [13], like fuzzy systems [14–18], neural networks [19–22], or even cooperative optimization [23].

Genetic algorithms are one of the most common heuristic methods of parameter tuning based on the minimization of computed or measured error signals. In practice, a dependence of the returned solutions on the particular specifications of the GA procedure (such as population size, mutation rate and so on) is observed, which enhances the need for better calibration of the GA parameters. These efforts can be categorized in three approaches [24]: (a) deriving the optimal values of a solution from empirical experiments, (b) controlling the GA parameters by varying them according to how often improvement of the current solution has been registered or including the GA running parameters in the chromosome codification, and (c) dimensional analysis and theoretical modeling. Also, the premature convergence of GAs has been another issue that has been discussed [25], together with the effects of non-random initial populations [26], and even genomic diversity [27, 28].

In order to overcome the shortcomings of the classic GAs [29–34] in tuning PID parameters, some applications have used a combination of genetic and interval algorithms [35], GA and immune system mechanism which introduce some recognition, immune memory and antibodies self-adjusting functions for main steam temperature control system [36], self-organizing genetic algorithms [37], adaptive genetic algorithms [38]. Another interesting approach regarding the increase of stabilization of a rotational inverted pendulum uses a variant of multi-objective genetic algorithms called Global Criterion Genetic Algorithm [39]. The good performances of the classic approach for tuning PID parameters with GAs is presented in [40], where a comparison with a variable parameter nonlinear PID (NL-PID) is made.

GAs are heuristic search algorithms based on the principles of natural selection. Their basic concept is designed to simulate processes in natural systems necessary for evolution. Their capabilities to find high performance solutions in large domains make them the ideal choice for optimization problems. However, GAs do not always guarantee an optimal solution. A new approach of the classic GA is presented in [41], in which the authors use the immune operator. Using this improved GA, they tune a set of PID parameters. The classic implementation of a PID controller is developed in [42], by analyzing the performance with mean square error (MSE) and integral absolute error (IAE) indexes using a classic GA and differential evolution (DE). An interesting approach is presented in [43] where the authors presents a new methodology for PID control tuning by coupling the Gain-Phase Margin method with GAs in which the micro population concept and adaptive mutation probability are applied.

Similar to finding a set of parameters specifically for controller design, GAs are also used in finding model parameters [44]. In [45] an improved version of the simple GA is presented, where the weights of the cost function are not taken as constant values, but varied throughout the procedure for parameter identification. This modified version of GA is applied to an induction motor as an example of nonlinear system. A consistent implementation using GAs is presented in [46], where authors have developed a model identification and controller design procedure for temperature control.

Some difficulties in the research for this type of problems using genetic algorithms are related to the implementation aspect. There are few frameworks and toolboxes that can help researchers continue existing work. A useful toolbox that implements genetic algorithms is presented in [47]. The toolbox is developed under MATLAB and is tested on a series of non-linear, multi-modal, non-convex test problems. In [48], numerical experiments based on heuristic approaches are conducted in order to reveal the characteristics of certain models, and then improved using artificial bee colonies and GAs. The proposed heuristic optimization algorithms were implemented in Microsoft Visual C++ using the .NET Framework. An interesting implementation designed for the modeling and simulation of Earth population evolution, on a global scale, throughout multiple historical eras is presented in [27]. The development of this software application has been made using JGAP [49], a genetic algorithms and genetic programming component provided as a Java framework.

This chapter brings forth the practical aspect of using genetic algorithms in aiding plant modeling and PID control design for real world industrial processes. Some plants are usually hard to model or tune on-site, especially after prolonged use of the equipment when degradation of performances is inevitable. Along with this presentation, a toolbox is offered (GAOT-ECM) in order to help researchers model processes, and (re)design viable PID controllers for their plants.

2 Principles of Evolutionary Computing

2.1 Evolutionary Computing Concepts

Evolutionary computing fits into the intelligent systems core idea of developing (more or less) intelligent computer systems by implementing simple rules for simple entities whose interaction will derive complex behaviors. In this case, complexity is achieved through artificial evolution, with application to search methods for problem solving. The possible solutions to a considered problem are coded into a population, which then is artificially evolved through the core mechanisms of selection, mutation and recombination.

Evolutionary algorithms (EAs) are heuristic search methods suitable for a wide range of optimization problems, specifically those that deal with high uncertainties and gross approximations, strict constraints or multiple criteria, vast solution spaces or high computational requirements.

Some of the more known categories of the evolutionary computing field are classified by the way the population is created. In genetic algorithms (GAs) the population represents potential solutions of the problem, being suitable for evolution through both mutation and crossover. For genetic programming (GP), the population represents potential programs, or procedures that can help solve a problem, whereas in evolutionary programming (EP) the population is formed from finite state automata. Other categories include evolution strategies (ES), gene expression programming (GEP), differential evolution (DE), and so on.

Evolution is controlled through a performance evaluation function which measures the adequacy/fitness of each individual to the environment to which it belongs. The steps for solving a problem using evolutionary computing are:

- Setting the coding mode of individuals from a population
- Choosing a method for generating a new population
- Setting the evaluation method for each individual
- Choosing the evolution operators and associated parameters

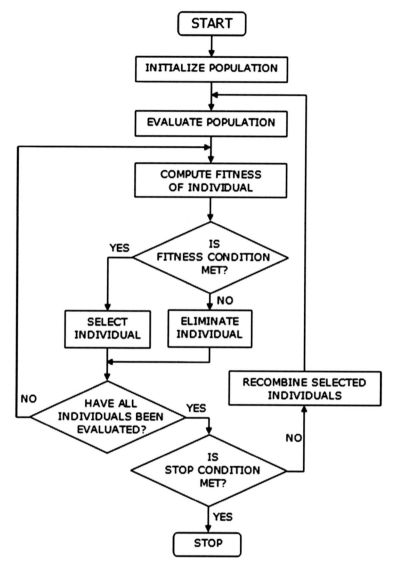

Fig. 1 Evolutionary algorithm steps

To run an EA, a series of steps are followed (Fig. 1): (1) the solution population is initialized, along with its fitness evaluation function; (2) as long as a previously declared termination condition is not met, for each generation, based on their computed fitness, some individuals are selected and recombined to generate offspring; and (3) the fitness of the current population is evaluated to be used in the next cycle.

In order to find the solution amongst a given population, EAs perform parallel searches on multiple directions by evaluating the adequacy of each individual, as a fitness variable, computed using a criterion or index that models a desired objective. Moreover, multi-criteria searches can be performed in order to find a desired solution amongst a given population of possible solutions. Used to model the general objective of the procedure, the description of the fitness function is problem-specific, whereas its purpose is to eliminate inadequate solutions from the gene pool. This function returns the value of the performance criteria (for example, a minimization index) which in turn will be used for selecting the parents of the new generation. The performance of the entire algorithm is given by the fitness of the best population during its run, or by the fitness of the final solution returned.

In order to obtain viable models and/or control laws for the considered plants and industrial processes, there are various selection mechanisms for the evaluated population. For instance, the roulette wheel selection generates a roulette wheel in which each individual is proportionally represented according to its fitness. The larger the fitness value, the more space it will occupy on the wheel. Thus, the individuals with higher fitness are more likely to be selected. The ranking selection mechanism uses a fitness-based sorted population list, from which the extreme values have been eliminated. This method has a slower convergence than roulette wheel selection. Tournament selection involves running several competitions or tournaments among a specified number of individuals randomly chosen from the population that is evaluated. The winner of each tournament is considered to be the one with the best fitness value, thus being selected for the subsequent recombination.

2.2 Genetic Algorithms

Genetic algorithms (GAs) are a subclass of evolutionary computing and are stochastic search algorithms inspired by natural selection and genetics. The principle of the GA is: given a population formed of individuals (chromosomes) that represent possible solutions to a problem, they are set to compete with each other for survival. Through evaluation, each individual is assigned a fitness value (describing its adequacy as a possible problem solution). The more fit ones are given a higher chance to participate in the reproductive process (through crossover and mutation), thus propagating more suitable solutions into the next generation, where, the resulting children will compete with each other and, possibly, also with their parents. In time, the population will be increasingly closer to the desired problem solution.

For GAs, the preparation steps are specific to the field of the problem, and are set by the user before launching into execution. The execution steps are independent of the problem itself, being executed by the computer that runs the procedures chosen for solving the problem.

The preparation steps must be specified by the user and consist of:

1. The set of independent, global, constant variables
2. Operators and their type
3. The fitness function
4. Execution parameters
5. Termination criteria

The execution steps are:

- Generation of the initial population
- Sequential execution of the next sub-steps until the termination criteria is met:

    - Creating a new population by applying two primary operations:

        Copying of existing individuals into the new population
        Creation of new individuals through the recombination (crossover) of certain parts from two individuals

    - Evaluating each individual from the new population and association of a fitness value depending on how well suited it is for solving the problem
    - Selecting the best individuals from the population.

- Selection of one, best individual from the population that represents the solution to the problem.

The fitness function provides the means to determine if an individual of the population is a better candidate than another, assigning a numeric value (the fitness value) that reflects the extent to which an individual meets the requirements of the problem at hand. In other words, this function is used to generate a partial order of candidates in the population, which is used in the execution steps of genetic recombination. A fitness function for a real world problem usually has multiple goals. This means that there could be more than one element to consider in the choice of fitness value assignment for an individual compared to others in a population.

The operators used in the development of genetic algorithms are: selection, crossover, and mutation. Depending on the nature of the problem, there are other operators that can be used (inversion/reversal, reordering, special operators). The most commonly used GA operators are:

- The operator of **selection**, based on fitness value
- The genetic operator of **crossover**
- The **mutation** operator, applied to a small number of individuals, either randomly or through a probability distribution function

Selection

An important role in designing a GA belongs to the selection operator. This operator decides which individuals are allowed to participate in the creation of the new population. The purpose of selection is to ensure greater chances of reproduction to selected individuals in a given population, thus seeking a maximization of their performance. Based on its purpose, this operator can be normalizing (this selection eliminates individuals with extreme fitness values), directional (a type of selection that aims to increase or decrease the average fitness of the entire population), or segregating (eliminates individuals with average fitness values). Many of the GAs implemented in practice use directional selection because it aims to achieve an optimal population. In this case, the difficulty consists in maintaining diversity inside a population, a task that is being left to the others operators. However, segregation is also used, on the premise that an individual with a very low fitness can become a potential solution at the next iteration, after crossover and mutation.

Selection can have zero tolerance, when only individuals identical with the current best are allowed to survive, it can have infinite tolerance, in which case all individuals survive regardless of their fitness, or a combination of both.

There are a few types of selection, from which some of the most popular are roulette, tournament, and ranking.

Roulette selection implies that individuals are chosen for recombination proportionally with their fitness value. Thus, the chromosomes with good fitness have better chances to spread their characteristics in future populations.

The tournament is one of the most popular and efficient schemes of selection, because, although very simple, it offers good performance in practice. It operates by randomly choosing individuals who will compete in a simulated game (by comparing fitness for example). The best of them will be selected for crossover.

Ranking selection is similar to roulette, only that it takes into account the presence of vastly different orders of magnitude for fitness. For example, if the fitness of one chromosome occupies 90 % of the roulette wheel, then the other chromosomes will have small chances of being selected. Therefore, ranking selection orders the individuals based on fitness and assigns a selection probability to each based on their place in the list.

Crossover

The crossover operator mimics natural inter-chromosomal recombination and is applied on individuals from intermediary generations. Two individuals are chosen randomly from the intermediary generation (which is also called crossover group/lot, resulted from selection), then certain parts of these individuals are interchanged. Most commonly, crossover operators of type (2, 2) are used, meaning two parents generate two descendants. Crossover exchanges information between the two parents, propagating their characteristics to the descendants. Crossover can be applied in one or two points, can be uniform, arithmetic or cut-and-splice, and so on.

One point crossover is at the core of recombination for GAs (Fig. 2). First, two chromosomes are chosen from the intermediary population. Second, they are split into the left and right chromosome parts, the cut being made at one random point. Third, each of the children receives the left chromosome from one of the parents and the right one from the other.

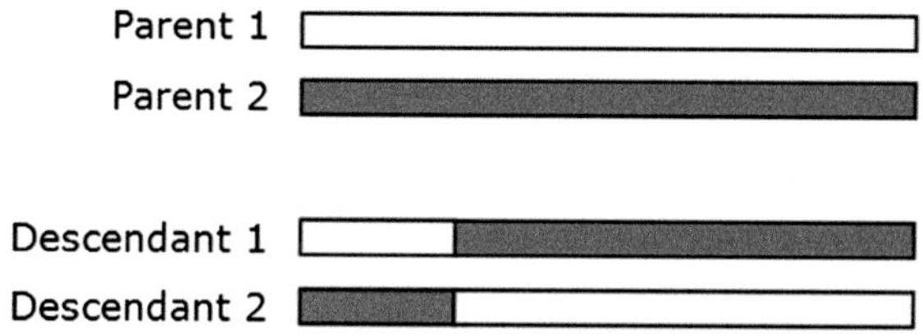

Fig. 2 One point crossover

Two point crossover is similar, the only difference being that two random cutting points are chosen. In uniform crossover, the chromosome genes (for instance, the bits for the binary representation) are interchanged randomly between the two parents according to an uniform probability distribution. In arithmetic crossover, a boolean or arithmetical operation is applied to the parents (for example, a logic AND in binary representation). The cut and splice method leads to changes of the resulting children lengths. This type of crossover is used when the chromosome codification has variable length.

Mutation

Through mutation, individuals who could not be obtained by other mechanisms are introduced into the population. The mutation operator's role is to maintain population diversity, and it acts on genes regardless of their position in the chromosome, by altering their value. Any gene of the chromosome can mutate. Mutation is a highly sensitive probabilistic operator that is applied according to a mutation rate chosen by the user. This value should be tuned, as a small mutation rate cannot maintain diversity, while a high mutation rate prevents or slows down the convergence to the optimal solution.

2.3   The Genetic Algorithm Optimization Toolbox Extension for Control and Modeling (GAOT-ECM)

In order to show how genetic algorithms can be applied to model identification and controller design, a toolbox was created, GAOT-ECM (an extension of the GAOT toolbox for control and modeling available for download at [50]). Both GAOT [47] and the extension presented GAOT-ECM are developed under MATLAB. GAOT is a good example of implementation of a genetic algorithm that can be used in many optimization problems.

From the types of selection presented in Sect. 2.2, this toolbox implements the roulette wheel, tournament and ranking methods. From the genetic operators (used to create new individuals inside populations) the toolbox implements some types of crossover and mutation. Both crossover and mutation operators are classified and implemented for float and binary representations of chromosomes, float and order-based representation. The types of crossover methods implemented in the GAOT package are: single and two points cut, arithmetic, uniform, cyclic (takes two parents P1 and P2 and performs cyclic crossover on permutation strings), linear order and order based. There are several types of mutation methods implemented in GAOT, from which the most popular are binary, uniform, boundary (randomly selects one gene and sets it equal to either its lower or upper bound), non-uniform (randomly selects one gene and sets it equal to an non-uniform random number), multi-non-uniform (applies the non-uniform method to all of the genes in the parent chromosome).

Beside the selection of the operators and their type, the implementation of a genetic algorithm also includes the initialization, termination, and evaluation functions. As mentioned in Fig. 1, an initial population must be provided before running the genetic algorithm. The GA iterates from generation to generation until a termination criterion is met. The most frequently used criterion for termination is a specified number of generations. Another condition refers to the convergence of population. The termination methods implemented in GAOT are: (a) reaching the maximum number of generations, and (b) reaching the maximum number of generation or finding a specified optimal fitness value. The evaluation functions can be implemented custom depending on the optimization problem. More details about the implementation of the GAOT toolbox are presented in [47].

Based on the GAOT toolbox, we have implemented an extension for this chapter in order to show how GAs can be used in model identification and controller design for plants like water tank, air blower, and motor. Another system on which GAOT-ECM is applied is the magnetorheological damper [51]. GAOT-ECM is implemented in a modular manner, therefore, depending on the user's level of experience, to be easy to configure and test. For more details about the GAOT-ECM implementation see Sect. 3.3.

3 Application of Genetic Algorithms to Model Identification and Controller Design

3.1 Plant Model Identification

In what concerns system identification with the aid of GAs, a structure of the model is chosen based on experimental data collected from the plant. The basic principle behind this method (Fig. 3) requires computing an approximation error ε between the model output y_m and the plant output y_{exp} (with the same system/model input u), which will be then used by the GA alongside its performance criterion I in order to find the best fitting model for the experimental data y_{exp}. The output of the GA is a

Fig. 3 Model identification
using genetic algorithms

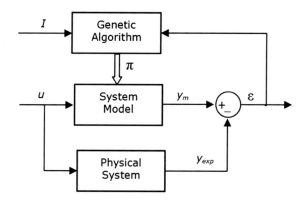

vector of model parameters π that are coded into chromosomes as potential problem
solutions.

Three performance criteria based on approximation error have been chosen for
this study (Table 1). A detailed description of these three indexes is given in
Sect. 3.2.

Step response plant analysis for linear time invariant models

In plant identification, **static analysis** is first used to obtain the domain for which
the plant behavior is modeled. For linear time systems with time invariant
parameters, the plant has approximately the same behavior for the entire interval in
which its static characteristic is linear. Using this tool, a nominal operating point
can be identified for the plant. The static characteristic $y(u)$ is obtained by gradually
increasing the plant input u within its admissible domain, and then reading the
measured output y in steady-state.

A series of **nominal operating points** can be chosen, usually toward the middle
of the linearity interval, around which there is high confidence that the plant
behavior will remain within the description of the model.

For the **dynamic analysis** of the plant, a step input around the nominal operating
point can be chosen, with an initial value u_0, and a final value u_1. The response of
the system to this step input will yield the y_{exp} experimental plant output that will be
used by the GA for identification.

Table 1 Performance criteria

ISE (integral of square error)	IAE (integral of absolute error)	ITAE (integral of time multiplied by the absolute value of error)
$I_1 = \int\limits_0^\infty \varepsilon^2(t)dt$	$I_2 = \int\limits_0^\infty |\varepsilon(t)|dt$	$I_3 = \int\limits_0^\infty t|\varepsilon(t)|dt$

In order to perform system identification with a linear time invariant transfer function, the experimental output data y_{exp} needs to be normalized:

1. convert data to percentages (if necessary)
2. shift data vertically so response starts in zero
3. scale data by input step size $u_1 - u_0$

This data will describe how the real system would respond to an ideal 0 to 1 step, if such an input were possible to be applied.

A real challenge is choosing the search domain. Caution is recommended in choosing the parameter bounds, through dynamic analysis: look at output steady state mean value for gain factor and settling time for time constants.

GA performance analysis

In order to quantify the performance of the GAOT-ECM tool, a performance index has been defined for the genetic algorithm:

$$J = \frac{1}{t} \int_0^t \varepsilon^2(t) dt \tag{1}$$

for a simulation time of t and an approximation error $\varepsilon(t)$.

Beside this index, a series of other parameters have been chosen to evaluate the genetic algorithm, such as run time and number of generations.

3.2 Controller Design

For the problem of controller design using GAs, the concept of the method, illustrated in Fig. 4, uses a chosen controller structure (in this case, a PID class controller) set in closed loop with either the model, in offline design, or the actual plant, in an online setting. Based on the control deviation ε (the error between setpoint r and controlled variable y), the genetic optimization technique will

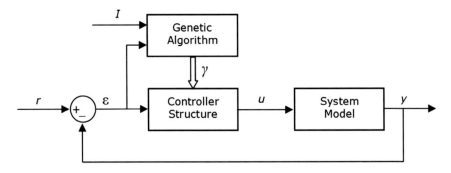

Fig. 4 Controller design using genetic algorithms

minimize a performance criterion I in order to find the best suited controller for the system at hand (whereas u is the command signal). The output of the GA is a vector of controller parameters γ that are coded into chromosomes as potential solutions of the controller design problem.

Thus, through this method, the parameters of the chosen PID controllers can also be fine-tuned in a simulated environment and then tested on the real world plant. There are a few performance indexes that are commonly used in the design process, each with their own advantages and disadvantages.

The ISE (Integral of Square Error) performance criterion is a general error minimizing index, making use of the quadratic error values obtained at a step response simulation. The main disadvantage of this index is that it does not take into account the overshoot. Instead, it just follows to quickly reduce the error values, thus reducing the overall settling time. In order to obtain low overshoot values, an addition to the fitness evaluation function of the GA is necessary.

$$ISE = \int_0^\infty \varepsilon^2(t)dt \qquad (2)$$

The IAE (Integral of Absolute Error) performance criterion is a general error minimizing index, making use of the absolute error values obtained at a step response simulation. The second performance index presents the same disadvantages as the first, but requires fewer processor operations, as instead of computing the square of error, only a removal of the sign is performed in order to obtain the absolute values.

$$IAE = \int_0^\infty |\varepsilon(t)|dt \qquad (3)$$

The ITAE (Integral of Time multiplied by the Absolute value of Error) performance criterion is one that minimizes the settling time, by using the absolute value of error obtained at a step response simulation and the actual value of time. The main disadvantage of this performance index is its high sensibility to any parameter variation, even if it produces lower overshoots that the first two indexes. Another issue of ITAE is the use of time values, which, for numeric systems, are subjected to processor clock communication delays.

$$ITAE = \int_0^\infty t|\varepsilon(t)|dt \qquad (4)$$

The IEC (Integral of Error and Command) performance criterion makes use of the command values in order to ensure minimization of the overshoot. This index successfully eliminates the overshoot. However, an increase in settling time is expected. Also, IEC requires more computing capacity than the first three indexes and is subject to the choice of the penalty factor ρ (strict positive).

$$IEC = \int_0^\infty \left(\varepsilon^2(t) + \rho u^2(t) \right) dt \tag{5}$$

Custom performance criteria can also be chosen. For step response performance, the most common requirements deal with overshoot (usually it is desired to be none at all or very small), settling time (a maximum is commonly set, but realistic results require setting a minimum also), rise time, and of course zero steady state error.

Choosing the controller parameter bounds is a challenge in this case as well. For example, when dealing with first order closed loop requirements and a first order plant, the integral time constant of the controller would be of the same size as the plant time constant. There are no recipes to be followed in this case. Usually, an experienced engineer would be able to estimate the size of the controller parameters and then manually tune them (going by the usual rules), but for higher order plants this initial estimation becomes less and less reliable. In the fortunate case that the plant model is extremely well identified, the theoretical methods (pole placement, frequency response, etc.) might work. It is however the norm, in industry, that the plant models are inaccurate due to reasons ranging from measurement noise to strict input restrictions.

3.3 Implementation

Both methods for model identification and controller design are implemented in the GAOT-ECM package. The toolbox is organized as follows: (1) identification and (2) control. A content overview file is provided so the users can easily identify the files that are necessary to test or customize it to their own problem. Depending on the users' experience, the configuration of this toolbox, i.e. the configuration of the proposed GA, was designed on three levels:

1. **Level 1** (for inexperienced users): represents the configurations that can be made inside the demonstration scripts (*ECMidentification* and *ECMcontrol* files); users can modify fitness function, plant model, performance criteria.
2. **Level 2**: represents some more advanced configurations that can be made in the main functions for control and modeling (*GAOT_ECM_ModelIdentification* and *GAOT_ECM_ControllerDesign* files); the parameters that can be modified refer to population size, optimal fitness value based on which the GA can end the execution.
3. **Level 3**: refers to the configuration that can be made regarding the GA, like number of generation, accepted tolerance, selection methods, crossover and mutation operators; these parameters can be changed inside the GA functions (*GAOT_ECM_ModelIdentification_GA*, *GAOT_ECM_ControllerDesign_GA*).

3.3.1 Model Identification

To demonstrate the usage of this functionality, a script is offered as an example (*ECMidentification.m*) that calls the main function that deals with model identification (*GAOT_ECM_ModelIdentification.m*). In order to run the main function some parameters must be set before calling it. These parameters represent the first level of configuration of the GA. If the user doesn't have any experience with genetic algorithms, and doesn't know how the GA could be tuned best, they should stop at this level.

Input parameters that must be set at the first level of configuration:

- Simulink model (*.mdl file) of the plant whose parameters need to be identified; GAOT-ECM provides plant models for: water tank, motor, and air blower. *Note*: GAOT-ECM provides all the necessary parameters and variables for the above models (detailed in the included case studies of this chapter). Regardless of the chosen model, these function are called and configured similarly.
- A *.mat file that is used in the identification process; it must contain the following variables:

  - Initial and general bounds for each parameter of the plant model (named *initBounds/generalBounds*).
  - Array containing the input and output data obtained experimentally using the hardware platform (named *in/out*).

- Fitness function (*.m file).
- Desired performance criteria used in computing the fitness function (the provided performance criteria are described in Table 1 from Sect. 3.1).

Output parameters:

- *bestIndividual*: the best solution found during the course of the run
- *endPop*: the final population
- *bPop*: a trace of the best population
- *PerformanceIndex*: quantifies the performance of the obtained model by computing the integral of square approximation error over simulation time
- *NumberOfGenerations*: a scalar value that represents the number of generations in which the GA has found the best solution
- *Fitness*: a scalar value that represents the fitness value of the best individual found

Based on the above configuration, the main function is called as presented below (having the input parameters defined as shown above):

```
[bestIndividual, endPop, bPop, PerformanceIndex,
  NumberOfGenerations, Fitness] = ...
  GAOT_ECM_ModelIdentification(criteria, fitnessFunction, ...
    mdlFileNameToUse, matFileToUse);
```

Inside the main function lies the second configuration level. More experienced users can modify this function in order to change some of the GA parameters (termination conditions, population size). In this function, the main GA function is called (*GAOT_ECM_ModelIdentification_GA.m*), and an analysis of the algorithm output is realized.

Input parameters for the GA main function:

- *criteria*: a scalar value that represents the user's choice regarding the performance criteria they want to use.
- *evalFN*: the name of the *.m file that represents the fitness function.
- *termFNOptimalValue*: a scalar value representing a constraint for the termination of the GA before a number of generations could be reached.
- *initBounds*: an array containing the initial bounds for each parameter of the model; each line of the array represents the bounds for one parameter.
- *varBounds*: an array containing the general bounds for each parameter of the model; each line of the array represents the bounds for one parameter.
- *populationSize*: a scalar value representing the number of individuals per generation.

The output parameters are almost the same with those of the main function. Based on the output, another simulation of the model is realized, with the best individual (the simulation time is another parameter that can be set at this configuration level) and two figures are generated: the model output using the proposed genetic algorithm and the convergence of the GA.

Based on the specified input and output parameters, the main GA can be called:

```
[bestIndividual, endPop, bPop, traceInfo] = ...
    GAOT_ECM_ModelIdentification_GA(criteria, evalFN, ...
    termFNOptimalValue, initBounds, varBounds, populationSize);
```

From the body of the GA, three steps can be identified (this is the deeper level of configuration):

- Initializing the first population (users can choose the number of generations, a tolerance value and other display options).
- Setting essential parameters for the GA (the user can choose from the options that GAOT provides; for more details see Sect. 2 or [47]):

  - The function used to terminate the GA: default setting evaluates if the maximum number of generations or desired fitness value was reached (variable: *termFN*).
  - Selection function: default setting is normal selection (variable: *selectFN*).
  - Crossover operator: an arithmetic method is chosen as default (variable: *xOverFNs*).
  - Mutation operator: an uniform mutation method is applied as default (variable: *mutFNs*).

Table 2 Examples of settings for plant modeling

Model	Fitness function name	Simulink model name	Information file name	Optimal fitness value
Water tank	evalwatertank	identificationwatertank	identWaterTankData	100
Air blower	evalair	identificationairblower	identAirBlowerData	100
Motor	evalmotor	identificationmotor	identMotorData	100

- Iterating the GA (calling the function *ga.m* developed in GAOT [47]).

The fitness function has the same structure for all plants: given a certain individual, whose positions are occupied by the model parameters, the Simulink model is executed. Then, based on its output and the selected performance criterion, the fitness value of the evaluated model is computed.

Some specific settings for the implemented plant models, useful to the identification procedures in the case studies can be observed in Table 2.

3.3.2 Controller Design

To demonstrate the usage of this functionality, a script is offered as an example (*ECMcontrol.m*) that calls the main function that deals with PID (Proportional-Integral-Derivative) controller design (*GAOT_ECM_ControllerDesign.m*). In order to run the main function, some parameters must be set before calling it. The input parameters are those presented in the model identification procedure, to which the imposed performance conditions that the controller must fulfill are added:

- Simulink model (*.mdl file) of the controller whose parameters need to be identified. GAOT-ECM provides Simulink files for: water tank, motor, and air blower.
  Note: GAOT-ECM provides all the necessary parameters and variables for the above files. Regardless of the chosen model these function are called and configured similarly.
- A *.mat file that must contain the following variables:
  - Initial and general bounds for each parameter of the PID controller (named *initBounds/generalBounds*).
  - The values of the plant models (named K, T_1, T_2 with meaning to each plant of the case studies specifically, presented in Sects. 4, 5, and 6).

- Fitness function (*.m file).
- Desired performance criteria used in computing the fitness function (the provided performance criteria are described in Sect. 3.2).
- Maximum settling time (in seconds).
- Minimum settling time (in seconds).
- Overshoot (percentage).

Table 3 Specific setting for controller design in GAOT-ECM

Model	Fitness function name	Simulink model name	Information file name
Water tank	evalcontrolwatertank	controlwatertank	controlWaterTankData
Air blower	evalcontrolair	controlairblower	controlAirBlowerData
Motor	evalcontrolmotor	controlmotor	controlMotorData

Table 4 Required closed loop performance

Model	Maximum settling time (s)	Minimum settling time (s)	Overshoot (%)	Optimal fitness value
Water tank	100	80	0.1	10
Air blower	10	3	0.1	10
Motor	6	2	0.1	10

The configuration levels are split like in the case of model identification. Regardless of the configuration level, the parameters and variables that the user can modify are the same as in the first case. The difference between the main function for control and the one for model identification is given by the output analysis. Some additional parameters are computed (specific to controller behavior, like settling time, overshoot, and steady state error), based on which the system closed loop response is plotted. The GA parameters (population size, number of generations, selection, crossover, and mutation methods) are set like in the model identification case.

The fitness function has the same structure for all controllers: given a certain individual, whose genes are occupied by the controller parameters, the Simulink control structure is executed. Based on its output, selected performance criteria and imposed performance for the controller, the fitness value is computed.

Some specific settings for the implemented controllers, relevant to the design procedure in the case studies included in this chapter are shown in Tables 3 and 4.

4 Case Study 1: Water Tank Experimental Model and Level Control

This case study presents the experimental modeling of a water tank as a second order model, by collecting experimental data which is fitted over a transfer function using genetic algorithms. The plant has one pole significantly larger than the other, the dominant behavior of the tank obscuring the delay introduced by water traveling through pipes. This is the basic sort of plant in industrial applications, whose model is then used to design a PID class controller, using the previously described evolutionary methodology.

Platform setup

The water tank used in this case study is part of a FESTO [52] laboratory installation, dedicated to the study of automation processes. The platform is comprised of two water tanks, one (tank A) mounted higher than the other (tank B), connected through pipes. For the level control functionality (in tank A), the platform makes use of a water pump as actuator, an ultrasonic transducer and the possibility of varying the evacuation water flow as disturbance. The water pump is mounted at the tank B level, using it as a water source. Between the water pump and the entry point into tank A, there are 0.97 m of pipes through which the water must travel. The platform is connected to a specialized application through an RS232 communication port and allows two main behaviors:

- manual control: in which the operator sets a specific command and can visualize the measured plant output
- automatic control: in which the operator sets a specific setpoint and can visualize the computed command and the measured plant output

The available application emulates a standard industry interface, allowing for visualization and input of various parameters. However, the application only allows for setting the parameters of a PID class controller, as is the case in many industrial settings, Kr (the PID proportional gain), Ti (the PID integral time constant) and Td (the PID derivative time constant), in the form:

$$u(t) = Kr \cdot \left(\varepsilon(t) + \frac{1}{Ti} \cdot \int_0^t \varepsilon(x)dx + Td \cdot \frac{d\varepsilon(t)}{dt} \right) \qquad (6)$$

where $u(t)$ is the command and $\varepsilon(t)$ is the control deviation.

Admissible command domains for actuators are dependant on physical characteristics (Table 5). In this case, the water pump has a lower saturation limit at 30 % command, while, due to the water fueling access point, maximum safe command is at 75 %, in order to avoid drawing air into the pump.

Static analysis

For this water tank, collected data yields the characteristic presented in Fig. 5. The linear domain of 40–70 % has been obtained by searching for a first degree polynomial fit. In this case, a nominal operating point has been chosen:

$$NOP_{tank} = (u_{NOP}, y_{NOP}) = (55, 18.34) \qquad (7)$$

Table 5 Technical characteristics

Component	Domain	Display domain on interface (%)	Admissible command domain
Level transducer	4–20 mA	0–100	Not applicable
Water pump	0–10 l/min	0–100	30–75 %

Fig. 5 Water tank static
characteristic

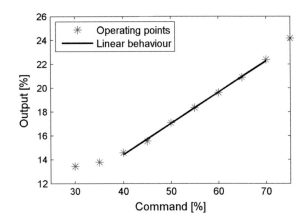

Dynamic analysis

For the dynamic analysis of the plant, a step input around the nominal operating point can be chosen, with an initial value u_0, and a final value u_1:

$$[u_0, u_1] = [40, 50][\%] \tag{8}$$

The measured output data has been collected over time. The normalized step response of the water tank is presented in Fig. 6. A very important observation is the high disturbance present in the measured output. Tank A of the system has a peculiarity in the form of a rubber stopper protruding from its wall inward, causing ripple effects on the surface of the water as it rises, which are being registered by the transducer. This effect is too severe to be eliminated from the data through a filtering mechanism. For its compensation, the tank would have to be replaced, which is not feasible. Therefore, when faced with this situation, caution is required when trying to estimate the plant model.

Upon initial inspection, this plant presents a first order behavior. However, there is also a small delay caused by water traveling from the pump to tank A through the existing pipe configuration, which can be either modeled as a time delay, or as a first

Fig. 6 Water tank
experimental step response

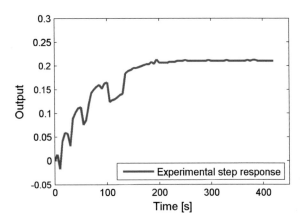

order delay. This characteristic is obscured by the dominant pole of the plant which yields a settling time of approximately 165 s (with a corresponding time constant of approximately 55 s). If a first order model is estimated and then a PI controller is designed, in all cases an overshoot of approximately 10 % can be observed each time. In these cases, the closed loop step response has the shape of the canonical step response of a second order function, which would not happen if the pipes were to introduce a time delay. Thus, a second order model with one dominant time constant T_1 significantly larger than the other one T_2 is chosen (with K as gain factor):

$$H_P(s) = \frac{K}{(T_1\,s + 1)(T_2\,s + 1)} \tag{9}$$

This sort of decision is deeply rooted in one's experience as an automation engineer. If the model chosen initially is not suitable for the considered plant, then it can be easily replaced with adding/removing poles or zeroes, and/or time delays.

Model identification with GAOT-ECM

Using the GAOT-ECM toolbox, the best model obtained for this plant is:

$$H_P(s) = \frac{K}{(T_1\,s + 1)(T_2\,s + 1)} = \frac{0.2151}{(66.1491\,s + 1)(4.6872\,s + 1)} \tag{10}$$

The model response fitted over experimental data is shown in Fig. 7. The criterion used for this solution is ITAE, while the performance index of the run is $J = 1.4327 \times 10^{-4}$, showing this is a very good approximation of the real plant.

Each GA run is affected by the input parameters of the algorithm, by the criterion chosen, and even by the computing system it runs on. In what follows, a series of comparative results are presented. All simulations have been run on a system equipped with an Intel Core i5 CPU @ 2.50 GHz and 4 GB of installed memory. Table 6 shows results for the three criteria ISE, IAE and ITAE. The GA is set to run for a population size of 100, and for a maximum of 100 generations. The selection

Fig. 7 Model and experimental responses

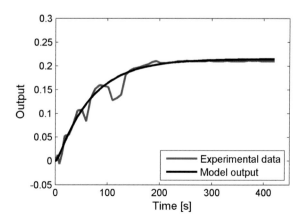

Table 6 Criterion variation: ISE, IAE, ITAE

Criterion	Run time (s)	Performance index J	Model parameters	Model fitness	Comparative response
ISE (I_1)	6.34 (9 generations)	1.5835×10^{-4}	K = 0.2155 $T_1 = 79.1836$ $T_2 = 3.3495$	$I_1 = 105.2522$	
IAE (I_2)	31.79 (100 generations)	1.4708×10^{-4}	K = 0.2197 $T_1 = 71.8339$ $T_2 = 4.6024$	$I_2 = 2.2297$	
ITAE (I_3)	43.38 (100 generations)	1.4327×10^{-4}	K = 0.2151 $T_1 = 66.1491$ $T_2 = 4.6872$	$I_3 = 21.7202$	

mechanism is ranking based on normalized geometric distribution, mutation is 8 % uniform, while the crossover operator is arithmetic. It is expected that the criteria values would be rather high, due to the disturbances observed in measured data. The most significant difference in results is, however, observed when using the ISE criterion, which yields a model that takes into account this output variation. For the other two criteria, the approximated models respond how the water tank should, if there were no constructive peculiarities, and thus no measured disturbance.

Table 7 presents the comparative performance of the genetic algorithm for various input parameters and settings, while using criterion ISE. As expected, a higher population size will yield good results faster, but requires more computing power. A too high mutation rate will degrade performances, as will a too relaxed stop condition. Restricting the search domain, naturally offers better results, but expanding it will yield various results, some better, some worse.

PID controller design with GAOT-ECM

Due to significant disturbances observed at the plant output, it is unwise to use a derivative component in the controller, as it has an unwanted amplification effect. Therefore, a PI controller is designed with parameters: Kr and $Ki = Kr/Ti$. For required performance of settling time between 80 and 100 s, no steady state error and no overshoot, the best controller found using GAOT-ECM is:

$$H_R(s) = Kr \cdot \left(1 + \frac{1}{Ti\ s}\right) = 9.9749 \cdot \left(1 + \frac{1}{70.2952\ s}\right) \qquad (11)$$

Table 7 Comparative GA performance

Settings				Outcome			
Population size	Mutation rate (%)	Stop at ISE	Search domain (K; T_1; T_2)	Number of generations	Run time (s)	Performance index J	Response shape
200	8	0.01	(0.01÷5; 1÷150; 0.1÷5)	3	6.87	1.3467×10^{-4}	G
100				6	5.11	1.5396×10^{-4}	S
50				64	23.05	1.6556×10^{-4}	S
20				15	6.97	1.6304×10^{-4}	S
100	2	0.01	(0.01÷5; 1÷150; 0.1÷5)	3	3.58	1.3688×10^{-4}	G
	12			6	5.94	1.4143×10^{-4}	G
	20			17	13.80	1.6943×10^{-4}	NS
100	8	0.02	(0.01÷5; 1÷150; 0.1÷5)	6	5.29	2.3878×10^{-4}	S
		0.05		2	4.14	1.7370×10^{-4}	NS
		0.1		2	3.52	12×10^{-4}	NS
100	8	0.01	(0.01÷8; 1÷200; 0.1÷8)	100	–	–	–
			(0.01÷1; 1÷100; 0.1÷2)	100	30.63	$1.\,3731 \times 10^{-4}$	G

NS non satisfactory, *S* satisfactory, *G* good

Fig. 8 Simulated closed loop
response

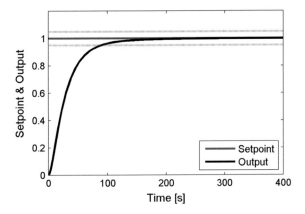

The platform application's interface only allows for one decimal place, therefore
forcing an approximation of the controller parameters:

- $Kr \approx 10$
- $Ti \approx 70.3$ s

Figures 8 and 9 show the performance of this controller, in simulation and
platform implementation. The performances of both cases are included in Table 8.

Table 9 shows results for the other four criteria ISE, IAE, ITAE, and ICE.
The GA is set to run for a population size of 100, and for a maximum of 100
generations. The selection mechanism is ranking based on normalized geometric
distribution, mutation is 8 % uniform, while the crossover operator is arithmetic.

As expected, using these criteria offers no real management of settling time and
overshoot. While steady state errors are indeed zero, the system will respond in
closed loop with overshoot, as the design process minimizes settling time without

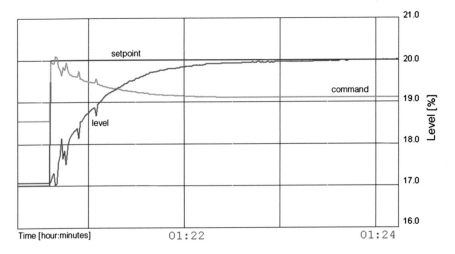

Fig. 9 Experimental closed loop response

Table 8 Controller performance

	Required performance	Simulated response	Experimental response
Steady state error	0	0	0
Settling time (s)	(80, 100)	85.92	82.268
Overshoot (%)	0.1	0	0

Table 9 Controller design with integral performance criteria

Criterion	Run time (s)	Steady state error (s)	Settling time (s)	Overshoot (%)	Controller parameters (Ki = Kr/Ti)	Closed loop response
ISE	29.13	0	112.75	20.4	Kr = 9.9901 Ki = 0.4492	
IAE	29.01	0	112.35	18.2	Kr = 9.9857 Ki = 0.4090	
ITAE	29.59	0	112.38	18.4	Kr = 9.9788 Ki = 0.4113	
ICE ρ = 0.006	30.15	0	199.86	0	Kr = 5.3351 Ki = 0.0706	

regard to a minimum of time the plant can realistically offer. These criteria, however, are extremely useful when searching for a closed loop response in which the setpoint is not a step, but a (randomly) variable signal.

5 Case Study 2: C.C. Motor Experimental Model and Speed Control

This case study presents the experimental modeling (and subsequent control system design) of a c.c. motor as a second order plant with both time constants of approximately the same order, which yields a particular time response of the system without overshoot. This sort of plant is difficult to model formally, as there is no viable correspondence between settling time of the response and the damping factor of the second order model when the latter is larger than one.

Platform setup

Experimental data is obtained using an in-house developed laboratory c.c. motor, dedicated to the study of motor speed control. The platform makes use of a digital encoder as transducer and a generator (AC to DC) to emulate the actuator (Table 10). The group transforms electrical energy received via the electrical grid into mechanical movement of the motor shaft. An RS232 communication port allows connection to a monitoring workstation.

The platform is connected to a specialized application implemented on an Intel MCS-51 microcontroller, and allows two main behaviors:

- manual control: in which the operator sets a specific command and can visualize the measured plant output
- automatic control: in which the operator sets a specific setpoint and can visualize the computed command and the measured plant output

The available application only allows for setting the parameters of a PID class controller, as is the case in many industrial settings, Kr (the PID proportional gain), Ki (the PID integral gain) and Kd (the PID derivative gain), in the form:

$$u(t) = Kr \cdot \varepsilon(t) + \frac{1}{Ki} \cdot \int_0^t \varepsilon(x)dx + Kd \cdot \frac{d\varepsilon(t)}{dt} \qquad (12)$$

where $u(t)$ is the command and $\varepsilon(t)$ is the control deviation.

Static analysis

For this plant, collected data yields the characteristic presented in Fig. 10. The linear domain of 30–90 % has been obtained by searching for a first degree polynomial fit. In this case, a nominal operating point has been chosen:

$$\text{NOP}_{\text{motor}} = (u_{\text{NOP}}, y_{\text{NOP}}) = (60.1, 1980) \qquad (13)$$

Dynamic analysis

For the dynamic analysis of the plant, a step input around the nominal operating point can be chosen, with an initial value u_0, and a final value u_1:

Table 10 Technical characteristics

Component	Domain	Display domain on interface	Admissible command domain
Transducer	0–12 V	800–3200 rpm	Not applicable
Actuator (%)	0.4–100	0.4–100	30–90

Fig. 10 Motor static
characteristic

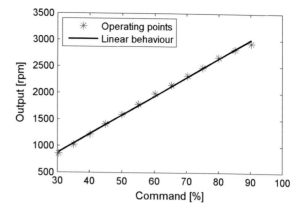

Fig. 11 Motor experimental
step response

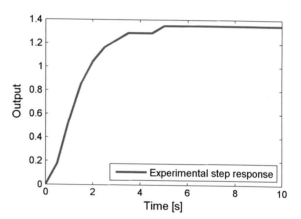

$$[u_0, u_1] = [55, 65]\,[\%] \tag{14}$$

The measured output data has been collected over time. The normalized step response of the motor is presented in Fig. 11.

Here, a very important initial observation is the right and downward inflexion of the response around the origin. If this were a first order plant, its response would curve left and upward. This sort of response belongs to second order plants for which the damping factor is higher than 1, whereas its poles are close. For these plants, estimation of time constants based on observed settling time has no theoretical foundation. Running GAOT-ECM to find these parameters is time efficient and more precise than trying to find the model through trial and error.

The chosen model is second order with two time constants T_1 and T_2 of the same order of magnitude, and gain factor K:

$$H_P(s) = \frac{K}{(T_1\,s + 1)(T_2\,s + 1)} \tag{15}$$

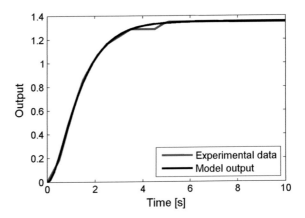

Fig. 12 Model and experimental responses

This sort of decision is deeply rooted in one's experience as an automation engineer. If the model chosen initially is not suitable for the considered plant, then it can be easily replaced with adding/removing poles or zeroes, and/or time delays.

Model identification with GAOT-ECM

Using the GAOT-ECM toolbox, the best model obtained for this plant is:

$$H_P(s) = \frac{K}{(T_1 s + 1)(T_2 s + 1)} = \frac{1.3519}{(0.8743 s + 1)(0.5678 s + 1)} \qquad (16)$$

The model response fitted over experimental data is shown in Fig. 12. The criterion used for this solution is ITAE, while the performance index of the run is $J = 1.6182 \times 10^{-4}$, showing this is a very good approximation of the real plant.

Each GA run is affected by the input parameters of the algorithm, by the criterion chosen, and even by the computing system it runs on. In what follows, a series of comparative results are presented. All simulations have been run on the same system as in the previous case. Table 11 shows results for the three criteria ISE, IAE and ITAE. The GA is set to run for a population size of 100, and for a maximum of 100 generations. The selection mechanism is ranking based on normalized geometric distribution, mutation is 8 % uniform, while the crossover operator is arithmetic.

Table 12 illustrates the comparative performance of the GA for several input parameters and settings, and criterion ISE. Results confirm the findings of Table 7, in terms of population size, mutation rate, search domain, and so on.

PID controller design with GAOT-ECM

A PID controller is designed with parameters: Kr, $1/Ki$, and Kd. For required performance of settling time between 2 and 10 s, no steady state error and no overshoot, the best controller found using GAOT-ECM is:

Table 11 Criterion variation: ISE, IAE, ITAE

Criterion	Run time (s)	Performance index J	Model parameters	Model fitness	Comparative response
ISE (I_1)	10.29 (20 generations)	1.5076×10^{-4}	$K = 1.3455$ $T_1 = 0.7901$ $T_2 = 0.6525$	$I_1 = 118.4488$	
IAE (I_2)	31.36 (100 generations)	1.4936×10^{-4}	$K = 1.3460$ $T_1 = 0.6400$ $T_2 = 0.7934$	$I_2 = 2.1729$	
ITAE (I_3)	31.24 (100 generations)	1.6182×10^{-4}	$K = 1.3519$ $T_1 = 0.8743$ $T_2 = 0.5678$	$I_3 = 0.0405$	

Table 12 Comparative GA performance

Settings				Outcome			
Population size	Mutation rate (%)	Stop at ISE	Search domain $(K; T_1; T_2)$	Number of generations	Run time (s)	Performance index J	Response shape
200	8	0.01	(0.1÷2; 0.1÷5; 0.1÷5)	4	6.91	1.5799×10^{-4}	G
100				20	10.29	1.5076×10^{-4}	G
50				52	18.05	1.7169×10^{-4}	G
20				78	31.39	1.6620×10^{-4}	G
100	2	0.01	(0.1÷2; 0.1÷5; 0.1÷5)	100	12.65	5.7368×10^{-4}	NS
	12			43	18.53	1.5744×10^{-4}	G
	20			100	61.61	4.5057×10^{-4}	NS
100	8	0.02	(0.1÷2; 0.1÷5; 0.1÷5)	5	4.04	3.16×10^{-4}	S
		0.05		4	4.16	4.2389×10^{-4}	NS
		0.1		2	3.07	7.6301×10^{-4}	NS
100	8	0.01	(0.1÷3; 0.1÷8; 0.1÷8)	100	–	–	–
			(1÷2; 0.1÷2; 0.1÷2)	38	13.85	1. 7454×10^{-4}	G

NS non satisfactory, *S* satisfactory, *G* good

Fig. 13 Simulated closed
loop response

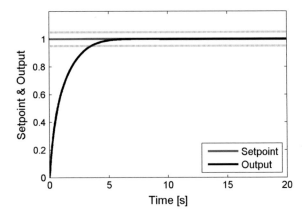

$$H_R(s) = Kr + \frac{1}{Ki\,s} + Kd\,s = 0.9994 + \frac{1}{0.6788\,s} + 0.4723\,s \qquad (17)$$

The platform application's interface only allows for one decimal place, therefore forcing an approximation of the controller parameters:

- $Kr \approx 1$
- $1/Ki \approx 1.5$ s
- $Kd \approx 0.4$ s

Figures 13 and 14 show the performance of this controller, in simulation and platform implementation. The performances of both cases are shown in Table 13.

Table 14 shows results for the other four criteria ISE, IAE, ITAE, and ICE. The GA is set to run for a population size of 100, and for a maximum of 100

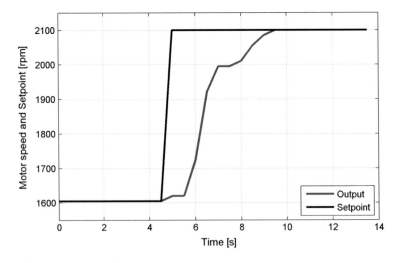

Fig. 14 Experimental closed loop response

Table 13 Controller performance

	Required performance	Simulated response	Experimental response
Steady state error	0	0	0
Settling time (s)	(2, 10)	3.51	∼4
Overshoot (%)	0.1	0	0

Table 14 Controller design with integral performance criteria

Criterion	Run time (s)	Steady state error (s)	Settling time (s)	Overshoot (%)	Controller parameters	Closed loop response
ISE	97.92	0	2.24	4.1	Kr = 0.9957 Ki = 0.9998 Kd = 0.4987	(plot: Setpoint & Output vs Time [s])
IAE	91.78	0	2.26	4.1	Kr = 0.9938 Ki = 0.9967 Kd = 0.4994	(plot: Setpoint & Output vs Time [s])
ITAE	85.48	0	2.25	4.2	Kr = 0.9901 Ki = 0.9977 Kd = 0.4997	(plot: Setpoint & Output vs Time [s])
ICE ρ = 0.2	183.75	0	3.34	0	Kr = 0.5801 Ki = 0.5279 Kd = 0.0851	(plot: Setpoint & Output vs Time [s])

generations. The selection mechanism is ranking based on normalized geometric distribution, mutation is 8 % uniform, while the crossover operator is arithmetic.

Here the run time is significantly higher than in the previous case study due to the presence of the derivative in the controller, which increases the evaluation time for each individual, as it requires more operations to compute. The results offered by the integral indexes are the same as in the previous case study.

6 Case Study 3: Air Blower Experimental Model and Control

This case study takes into account the presence of time delay for an air flow plant of first order. The plant model is obtained using experimental data from a laboratory air blower actuated by a fan. As in the other case studies, a PID class controller is obtained, this time taking account of the inherent time delay of the system, then validated in closed loop.

Platform setup

The ELWE LTR701 air blower [53] is a laboratory installation dedicated to the study of automation processes. The platform is comprised of a long tubular nozzle, in which air is drawn using a motor actuated fan (Table 15). The air pressure sensor is installed at the other end of the nozzle, thus introducing a time delay in the system, whereas the rest of the plant behavior is that of a first order system. The platform is connected to a specialized application through an RS232 communication port and allows two main behaviors:

- manual control: in which the operator sets a specific command and can visualize the measured plant output
- automatic control: in which the operator sets a specific setpoint and can visualize the computed command and the measured plant output

The available application emulates a standard industry interface, allowing for visualization and input of various parameters. However, the application only implements a PI controller for the air pressure functionality, with Kr (the PI proportional gain), and Ki (the PI integral gain), in the form:

$$u(t) = Kr \cdot \varepsilon(t) + Ki \cdot \int_0^t \varepsilon(x)dx \qquad (18)$$

where $u(t)$ is the command and $\varepsilon(t)$ is the control deviation.

Static analysis

For this plant, collected data yields the characteristic presented in Fig. 15. The linear domain has been estimated to 25-100 [%]. In this case, a nominal operating point has been chosen:

Table 15 Technical characteristics

Component	Domain (V)	Display domain on interface	Admissible command domain
Pressure sensor	0–10	0–1000 N/m^2	Not applicable
Fan	0–10	0–100 %	5–100 %

Fig. 15 Air blower static characteristic

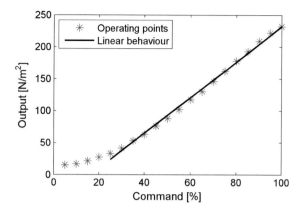

$$NOP_{tank} = (u_{NOP}, \; y_{NOP}) = (55, 102) \tag{19}$$

Dynamic analysis

For the dynamic analysis of the plant, a step input around the nominal operating point can be chosen, with an initial value u_0, and a final value u_1:

$$[u_0, u_1] = [45, 65] \, [\%] \tag{20}$$

The measured output data has been collected over time. The normalized step response of the air blower is presented in Fig. 16. A first order model with one time constant T_1 and a time delay T_2 is chosen, with K as gain factor:

$$H_P(s) = \frac{K}{T_1 \, s + 1} e^{-T_2 \, s} \tag{21}$$

This sort of decision is deeply rooted in one's experience as an automation engineer. If the model chosen initially is not suitable for the considered plant, then it can be easily replaced with adding/removing poles or zeroes, and/or time delays.

Fig. 16 Air blower experimental step response

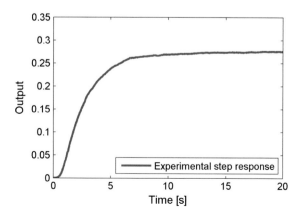

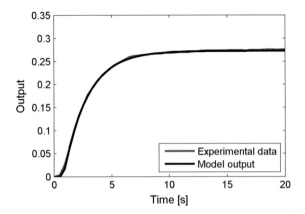

Fig. 17 Model and experimental responses

Model identification with GAOT-ECM

Using the GAOT-ECM toolbox, the best model obtained for this plant is:

$$H_P(s) = \frac{K}{T_1\,s + 1}\,e^{-T_2\,s} = \frac{0.2729}{2.0699\,s + 1}\,e^{-0.7754\,s} \qquad (22)$$

The model response fitted over experimental data is shown in Fig. 17. The criterion used for this solution is IAE, while the performance index of the algorithm run is $J = 3.7685 \times 10^{-6}$, showing this is a very good approximation of the real plant.

Each GA run is affected by the input parameters of the algorithm, by the criterion chosen, and even by the computing system it runs on. In what follows, a series of comparative results are presented. All simulations have been run on the same system as before. Table 16 shows results for the three criteria ISE, IAE and ITAE. The GA is set to run for a population size of 100, and for a maximum of 100 generations. The selection mechanism is ranking based on normalized geometric distribution, mutation is 8 % uniform, while the crossover operator is arithmetic.

Table 17 presents the comparative performance of the GA for various input parameters and settings (with criterion ISE). As expected, the results of Tables 7 and 12 are confirmed, amongst which: a too low mutation rate will degrade performances, whereas a higher mutation rate is beneficial. This happens due to the too fast convergence of the algorithm toward the stop condition, which can be also observed. Adding a stricter condition (for example request that a viable individual must have a value of ISE at most 0.002) will avoid the premature stop of the GA run. Restricting the search domain doesn't necessarily improve the model due to the fast convergence, but expanding it will yield both better and worse results.

PID controller design with GAOT-ECM

Due to significant disturbances observed at plant output, it is unwise to use a derivative component in the controller, as it has an unwanted amplification effect.

Table 16 Criterion variation: ISE, IAE, ITAE

Criterion	Run time (s)	Performance index J	Model parameters	Model fitness	Comparative response
ISE (I_1)	3.09 (2 generations)	1.2387×10^{-4}	K = 0.2753 T_1 = 2.4971 T_2 = 0.1808	I_1 = 146.7828	
IAE (I_2)	26.85 (100 generations)	3.7685×10^{-6}	K = 0.2729 T_1 = 2.0699 T_2 = 0.7754	I_2 = 0.0690	
ITAE (I_3)	29.42 (100 generations)	3.9411×10^{-6}	K = 0.2745 T_1 = 2.1877 T_2 = 0.6766	I_3 = 0.2497	

Therefore, a PI controller is designed with parameters: *Kr* and *Ki*. For required performance of settling time between 3 and 10 s, no steady state error and no overshoot, the best controller found using GAOT-ECM is:

$$H_R(s) = Kr + Ki\frac{1}{s} = 3.92 + 1.84\frac{1}{s} \tag{23}$$

The platform application's interface allows several decimal places, therefore no further approximation was made for the controller parameters.

Figures 18 and 19 show the performance of this controller, in simulation and platform implementation. The performances of both cases are shown in Table 18.

Table 19 shows results for the other four criteria ISE, IAE, ITAE, and ICE. The GA is set to run for a population size of 100, and for a maximum of 100 generations. The selection mechanism is ranking based on normalized geometric distribution, mutation is 8 % uniform, while the crossover operator is arithmetic.

Results confirm the conclusions drawn from the previous case studies regarding the use of integral criteria. However, even if the closed loop response is not entirely satisfactory, its shape can still be changed by manipulating the search domains for the controller parameters (for instance, increasing the integral factor will lower the overshoot, but increase the settling time). Here, it is also very important to note that minimizing deviation error will also lead invariably to closed loop stable control systems, the time delay of the plant notwithstanding.

Table 17 Comparative GA performance

Settings				Outcome			
Population size	Mutation rate (%)	Stop at ISE	Search domain (K; T_1; T_2)	Number of generations	Run time (s)	Performance index J	Response shape
200	8	0.01	$(0.01\div2; 0.1\div5; 0.01\div1)$	2	6.06	6.4625×10^{-6}	G
100				2	3.09	1.2387×10^{-4}	S
50				3	2.45	5.8799×10^{-5}	S
20				17	7.17	1.3309×10^{-4}	NS
100	2	0.01	$(0.01\div2; 0.1\div5; 0.01\div1)$	2	3.02	7.3658×10^{-5}	NS
	12			2	3.26	1.7446×10^{-4}	NS
	20			2	3.52	3.0038×10^{-5}	S
100	8	0.02	$(0.01\div2; 0.1\div5; 0.01\div1)$	2	3.24	8.1721×10^{-5}	NS
		0.05		2	3.13	1.2827×10^{-4}	NS
		0.1		2	3.16	4.0878×10^{-4}	NS
		0.002		6	4.84	1.2735×10^{-5}	G
100	8	0.002	$(0.01\div4; 0.1\div8; 0.01\div2)$	100	–	–	–
			$(0.01\div1; 1\div5; 0.05\div0.8)$	2	3.20	4.9998×10^{-6}	S

NS non satisfactory; *S* satisfactory; *G* good

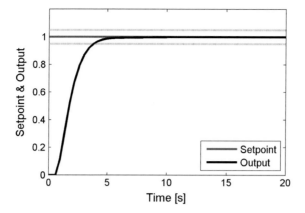

Fig. 18 Simulated closed loop response

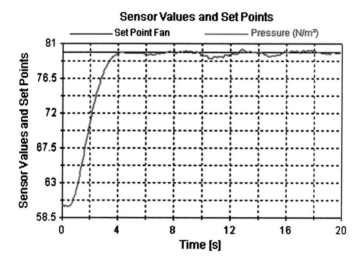

Fig. 19 Experimental closed loop response

Table 18 Controller performance

	Required performance	Simulated response	Experimental response
Steady state error	0	0	0
Settling time (s)	(3, 10)	3.4	3.4355
Overshoot (%)	0.1	0	0

Table 19 Controller design with integral performance criteria

Criterion	Run time (s)	Steady state error (s)	Settling time (s)	Overshoot (%)	Controller parameters	Closed loop response
ISE	32.02	0	9.4	6.20	Kr = 3.93 Ki = 2.45	
IAE	29.629	0	3.33	8.2	Kr = 5.87 Ki = 2.60	
ITAE	31.71	0	2.98	0.9	Kr = 4.34 Ki = 2.07	
ICE ρ = 0.01	32.47	0	7.19	0.4	Kr = 2.01 Ki = 1.14	

7 Case Study 4: Magnetorheological Damper Control

A magnetorheological (MR) damper is a hydraulic-class actuator used in seismic protection systems design [54]. This device is used to generate the necessary control forces using as an input a command current and the velocity of the story on which it is mounted. The damper is a hydraulic cylinder whose damping coefficient is controlled by the variation of a magnetic field which changes the fluid from viscous to semi-solid in milliseconds [55].

For the successful vibration control of a given structure, it is necessary that the actuators perform inside a set of strict performance criteria, such as response time and robustness versus uncertainties [56]. The output forces generated by seismic dampers are required to be maintained between specific limits, so they do not cause instability to the structure, breaks support beams and so on. Therefore, a control loop for the actuator is required, that will receive the desired control forces computed by higher algorithms (such as robust laws, intelligent controllers, adaptive, modal, etc.) as setpoints and ensure that they are precisely reproduced by the actual damper output force.

The behavior of the MR damper used in this case study is given by [57]:

$$
\begin{cases}
F = c_0 \dot{x} + \alpha z \\
\dot{z} = -\delta |\dot{x}| z |z|^{n-1} - \beta \dot{x} |z|^n + A \dot{x} \\
\alpha = \alpha_a + \alpha_b u \\
c_0 = c_{0a} + c_{0b} u \\
\dot{u} = -\eta(u - u_V)
\end{cases}
\tag{24}
$$

where F is the force generated by the damper, u is the command signal, u_V is the command voltage, $\dot{x}$ is the velocity of the structure. This case study makes use of the following values [58]: $c_{0a} = 0.0064$ Ns/cm, $c_{0b} = 0.0052$ Ns/cm V, $a_a = 8.66$ N/cm, $a_b = 8.86$ N/cm V, $\delta = 300$ cm^{-2}, $\beta = 300$ cm^{-2}, $A = 120$, $n = 2$ and $\eta = 80$ s^{-1}.

For the purpose of this case study, a base isolation system is considered for a three story building. The damper is mounted in the base of the structure, which is controlled via an LQR law on the outer loop of a cascaded control system. The design on the LQR law, the building model and its state space representation are described in [55]. In order to maintain the computational requirements of the control system to a minimum, a PID controller is chosen to be designed using GAs for the inner loop containing the damper. This configuration is presented in Fig. 20, where: x is the displacement of the structure; $\dot{x}$ and $\ddot{x}$ are the velocity and acceleration of the structure, respectively; F is the control force as damper output; u_{LQR} is the desired control force, u is the command signal, ε is the control deviation of the inner loop; a and v are the ground acceleration and velocity, respectively.

The PID damper controller is chosen to be of the form:

$$
u(t) = Kr \cdot \varepsilon(t) + Ki \cdot \int_0^t \varepsilon(x)dx + Kd \cdot \frac{d\varepsilon(t)}{dt}
\tag{25}
$$

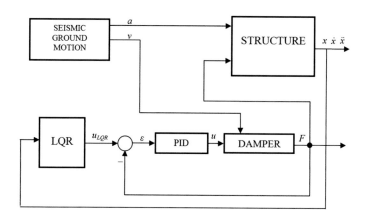

Fig. 20 Cascaded control structure for building equipped with MR damper

where $u(t)$ is the command, $\varepsilon(t)$ is the control deviation, Kr is the PID proportional gain, Ki the integral gain, and Kd the derivative gain.

For this system, the only relevant performance criteria are ISE, IAE, ITAE, and IEC, as the setpoints will never be step, but quick variations of a signal computed based on similarly quasi-random signals, like seismic ground motion. The purpose of the inner control loop is to follow the setpoint precisely. Thus, one of the best controller found using GAOT-ECM wit the ISE criterion is:

$$H_R(s) = Kr + Ki \cdot \frac{1}{s} + Kd \ s = 248.8 + 91.38 \cdot \frac{1}{s} + 29.69 \ s \qquad (26)$$

Figure 21 shows the PID controller performance for a 3 s timeframe: the force output is superimposed onto the given setpoint, as requested. The fitness of this solution is given by the criterion value: ISE = 0.0193.

In order to validate this control system, a look at the entire structural response is required. In what follows, a set of evaluation criteria [59] has been chosen:

$$\begin{cases} J_1 = \frac{\max |x_i(t)|}{x_{open}} \\ J_2 = \frac{\max |d_i(t)|}{d_{open}} \\ J_3 = \frac{\max |\ddot{x}_i(t)|}{\ddot{x}_{open}} \end{cases} \qquad (27)$$

where $x_i(t)$ and $\ddot{x}_i(t)$ are the relative displacement and acceleration of the i-th story, while $d_i(t)$ is the interstory drift; the notation *open* designates the overall maximum absolute displacements, accelerations and drifts of the uncontrolled structure.

Figure 22 presents the structural response (overall maximum displacements, interstory drifts, and accelerations) of the cascaded control system compared to both the standard non-cascaded control loop, and the uncontrolled structure when subjected to the Northridge (1994) earthquake. Since the LQR control law is computed using the linearized model of the structure, it doesn't take into account the nonlinear behaviour of the damper. Results show that without the inner PID control loop, the

Fig. 21 PID controller performance for a 3 s timeframe

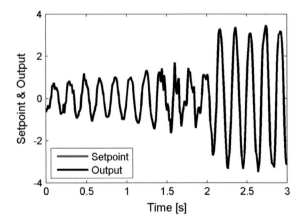

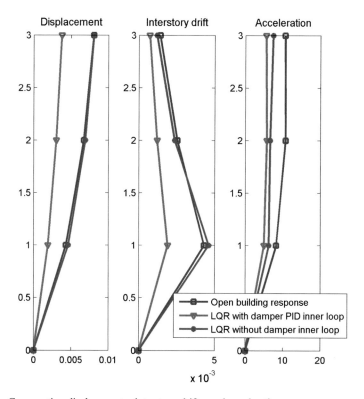

Fig. 22 Comparative displacements, interstory drifts, and accelerations

structural response is only improved in what concerns accelerations, even worsens when it comes to interstory drifts. The beneficial effects of the inner control loop are obvious in an overall considerable reduction of the structural response: 54.31 % reduction in displacements, 63.09 % reduction of interstory drifts, and 15.87 % reduction of structure accelerations.

Table 20 presents the results of GA runs using various configuration parameters and performance criteria. The expected closed loop performances for the structural system with MR damper base isolation are preserved. Naturally, due to computational requirements for running the nonlinear damper model, the run times are longer than in the previous case studies. For the ITAE criterion, expected fitness values are considerably larger than for the other three, due to its dependence to length of simulation. The fast convergence of the runs using criteria ISE and IAE is not premature, but obtained through large enough search domains and a carefully chosen stop condition. For requirements related to obtaining smaller admissible values for the PID controller parameters, the IEC criterion is the best choice, but it's highly sensitive to the value of the ρ factor.

An implementation of this case study is available for download at [51]. The file uses the GAOT-ECM extension described in Sect. 3.3. Along with the genetic

Table 20 Controller design with integral performance criteria

Criterion	Run time (s)	Number of generations	Solution fitness	Controller parameters	Cascaded loop response	No inner loop	Reduction (%)
ISE	24.57	2	0.0193	$Kr = 248.83$ $Ki = 91.38$ $Kd = 29.69$	$J_1 = 0.4634$ $J_2 = 0.4393$ $J_3 = 0.5311$	$J_1 = 1.0065$ $J_2 = 1.0702$ $J_3 = 0.6898$	54.31 63.09 15.87
IAE	22.69	2	1.3153	$Kr = 238.48$ $Ki = 141.89$ $Kd = 43.66$	$J_1 = 0.4633$ $J_2 = 0.4392$ $J_3 = 0.5312$	$J_1 = 1.0065$ $J_2 = 1.0702$ $J_3 = 0.6898$	54.32 63.10 15.86
ITAE	200.52	100	378.4544	$Kr = 248.79$ $Ki = 149.90$ $Kd = 44.02$	$J_1 = 0.4633$ $J_2 = 0.4392$ $J_3 = 0.5312$	$J_1 = 1.0065$ $J_2 = 1.0702$ $J_3 = 0.6898$	54.32 63.10 15.86
ICE $\rho = 0.001$	228.24	100	5.3173	$Kr = 176.42$ $Ki = 54.83$ $Kd = 0.02$	$J_1 = 0.4641$ $J_2 = 0.4399$ $J_3 = 0.5313$	$J_1 = 1.0065$ $J_2 = 1.0702$ $J_3 = 0.6898$	54.24 63.03 15.85

algorithm functions, the file includes the simulation models for the MR damper and the structure (open and with base isolation). Seismic data for the Northridge (1994) earthquake is provided, as well as a previously computed command matrix for the LQR law for validation purposes.

8 Conclusions

This chapter offers an applicative view of evolutionary computing, specifically genetic algorithms, in regards to the modeling and control of industrial plants. From measurement noise to time delays, from inflexible control interfaces to restrictions on system-wide reconfigurations, and even from equipment wear to unforeseeable uncertainties, practicing automation engineers are sometimes faced with difficult choices when it comes to tuning PID class controllers on-site, especially during system functioning. It is a running pun in the control community that "automation is the art of approximation": there is a seed of truth in this, as most practitioners rely heavily on personal experience when performing the final tuning of controllers during installation. This chapter comes to the aid of those practitioners who have not yet built this base of knowledge, or those who want to enrich their expertise.

Running a genetic algorithm equates with a multi-directional search in easily expandable or retractable parameter domains, for both models and controllers. Suitable controllers, or at least suitable parameter domains for on-site tuning, are hard to find. Genetic algorithms (GAs) offer this search quickly and safely, without damaging the plant. With possible implementations that can run in under a minute on a mobile device, GAs have become a useful tool in control engineering practice.

For this work, a MATLAB extension to the Genetic Algorithm Optimization Toolbox has been implemented, GAOT-ECM (Extension for Control and Modeling) and made available through MATLAB Central's File Exchange service. The four case studies included can be run and used in simulations, extended to other systems, or even used for educational purposes.

The four case studies include the modeling and control of a water tank, c.c. motor, an air blower, and a magnetorheological damper each with its own peculiarities that can be observed in real world industrial systems. After the successful plant identification and subsequent PID design (with validation on the physical system), a series of comparative results have been discussed. The usefulness of performance criteria is highly dependant on the problem at hand, while an increase in population size yields better solutions, but increases computing time. In what concerns modeling, a highly disturbed output will affect any identification algorithm, but the GA can offer both approximations that seem formally viable, and approximations that emulate the seemingly intuitive response tracing of an engineer's experienced hand. Moreover, GAs can find plant models that are not easily identified formally, for instance when a response is second order LTI, but over-damped. The same happens when a dominant time constant obscures a much smaller one, which would significantly affect the closed loop response of any formally designed controller.

Our parallel research directions have included successful GA model identification for human heart rate during aerobic training, GA search for PID controllers for semi-active seismic dampers (from the electrohydraulic and magnetorheological classes), and even GA tuning of multiple input multiple output PID controllers for nonlinear plants (a 3D crane). For the immediate future, we are looking at applying GA optimization to multi-agent control systems, to path reconfiguration in real world environments for emergency vehicles in high urban traffic, and last, but not least, to optimal damper distribution in tall buildings for mitigating seismic response.

In closing, we are issuing a challenge to our readers to add their own plants, linear or nonlinear, stable or not, simulated or experimental, to the long list of applications that GAs stand to offer.

References

1. Malhotra, R., Singh, N., Singh, Y.: Genetic algorithms: Concepts, design for optimization of process controllers. Comput. Inf. Sci. **4**(2), 39 (2011)
2. Mitchell, M.: An Introduction to Genetic Algorithms. MIT Press, Cambridge (1998)
3. Fleming, P.J., Purshouse, R.C.: Evolutionary algorithms in control systems engineering: a survey. Control Eng. Pr. **10**(11), 1223–1241 (2002)
4. Fonseca, C.M., Fleming, P.J.: Multiobjective genetic algorithms. In IEE Colloquium on Genetic Algorithms for Control Systems Engineering, pp. 1–6 (1993, May)
5. Lewin, D.R.: A genetic algorithm for MIMO feedback control system design. Adv. Control Chem. Process. **1994**, 101 (2014)
6. Lewin, D.R.: Multivariable feedforward control design using disturbance cost maps and a genetic algorithm. Comput. Chem. Eng. **20**(12), 1477–1489 (1996)
7. Acosta-González, E., Fernández-Rodríguez, F.: Model selection via genetic algorithms illustrated with cross-country growth data. Empir. Econ. **33**(2), 313–337 (2007)
8. Huang, C.F.: A hybrid stock selection model using genetic algorithms and support vector regression. Appl. Soft Comput. **12**(2), 807–818 (2012)
9. Gray, G.J., Murray-Smith, D.J., Li, Y., et al.: Nonlinear model structure identification using genetic programming. Control Eng. Pr. **6**(11), 1341–1352 (1998)
10. Bush BO, Hosom JP, Kain A et al.: Using a Genetic Algorithm to Estimate Parameters of a Coarticulation Model. In: INTERSPEECH, pp. 2677–2680 (2011)
11. Castiglione, A., Cattaneo, G., Cembalo, M., et al.: Experimentations with source camera identification and Online Social Networks. J. Ambient Intell. Humaniz. Comput. **4**(2), 265–274 (2013)
12. Vatolkin, I., Preuß, M., Rudolph, G., et al.: Multi-objective evolutionary feature selection for instrument recognition in polyphonic audio mixtures. Soft. Comput. **16**(12), 2027–2047 (2012)
13. De Santis, A., Castiglione, A., Fiore, U., et al.: An intelligent security architecture for distributed firewalling environments. J. Ambient Intell. Humaniz. Comput. **4**(2), 223–234 (2013)
14. Alcalá-Fdez, J., Alcalá, R., Gacto, M.J., et al.: Learning the membership function contexts for mining fuzzy association rules by using genetic algorithms. Fuzzy Sets Syst. **160**(7), 905–921 (2009)

15. Shook, D.A., Roschke, P.N., Lin, P.Y., et al.: GA-optimized fuzzy logic control of a large-scale building for seismic loads. Eng. Struct. **30**(2), 436–449 (2008)
16. Linkens, D.A., Nyongesa, H.O.: Genetic algorithms for fuzzy control. 1. Offline system development and application. IEE Proc.-Control Theor. Appl. **142**(3), 161–176 (1995)
17. Karr, C.L., Gentry, E.J.: Fuzzy control of pH using genetic algorithms. IEEE Trans. Fuzzy Syst **1**(1), 46 (1993)
18. Herrera, F., Lozano, M., Verdegay, J.L.: A learning process for fuzzy control rules using genetic algorithms. Fuzzy Sets Syst. **100**(1), 143–158 (1998)
19. Tao, Q., Liu, X., Xue, M.: A dynamic genetic algorithm based on continuous neural networks for a kind of non-convex optimization problems. Appl. Math. Comput. **150**(3), 11–820 (2004)
20. Javadi, A.A., Farmani, R., Tan, T.P.: A hybrid intelligent genetic algorithm. Adv. Eng. Inform. **19**(4), 255–262 (2005)
21. Leung, F.H., Lam, H.K., Ling, S.H., et al.: Tuning of the structure and parameters of a neural network using an improved genetic algorithm. IEEE Trans. Neural Netw. **14**(1), 79–88 (2003)
22. Schaffer, J.D., Whitley, D., Eshelman, L.J.: Combinations of genetic algorithms and neural networks: a survey of the state of the art. In: International Workshop on Combinations of Genetic Algorithms and Neural Networks, 1992., COGANN-92, pp. 1–37 (1992, June)
23. Zheng, Y.J., Ling, H.F.: Emergency transportation planning in disaster relief supply chain management: a cooperative fuzzy optimization approach. Soft. Comput. **17**(7), 1301–1314 (2013)
24. Gibbs, M.S., Dandy, G.C., Maier, H.R.: A genetic algorithm calibration method based on convergence due to genetic drift. Inf. Sci. **178**(14), 2857–2869 (2008)
25. Chang, P.C., Huang, W.H., Ting, C.J.: Dynamic diversity control in genetic algorithm for mining unsearched solution space in TSP problems. Expert Syst. Appl. **37**(3), 1863–1878 (2010)
26. Togan, V., Daloglu, A.T.: An improved genetic algorithm with initial population strategy and self-adaptive member grouping. Comput. Struct. **86**(11), 1204–1218 (2008)
27. Patrascu, M., Stancu, A.F., Pop, F.: HELGA: a heterogeneous encoding lifelike genetic algorithm for population evolution modeling and simulation. Soft. Comput. **18**(12), 2565–2576 (2014)
28. Lässig, J., Sudholt, D.: Design and analysis of migration in parallel evolutionary algorithms. Soft. Comput. **17**(7), 1121–1144 (2013)
29. Mitsukura, Y., Yamamoto, T., Kaneda, M.: A genetic tuning algorithm of PID parameters. In: IEEE International Conference on Systems, Man, and Cybernetics, 1997. Computational Cybernetics and Simulation, 1997, vol. 1, pp. 923–928 (1997, October)
30. Ding, Y.M., Wang, X.Y.: Real-coded adaptive genetic algorithm applied to PID parameter optimization on a 6R manipulators. In: Fourth International Conference on Natural Computation, 2008. ICNC'08, vol. 1, pp. 635–639 (2008, October)
31. Chen, Y., Wu, Q.: Design and implementation of PID controller based on FPGA and genetic algorithm. In: 2011 International Conference on Electronics and Optoelectronics (ICEOE), vol. 4, pp.4–308 (2011, July)
32. Juang, J.G., Huang, M.T., Liu, W.K.: PID control using presearched genetic algorithms for a MIMO system. IEEE Trans. Syst. Man Cybern. Part C: Appl. Rev. **38**(5), 716–727 (2008)
33. Valarmathi, R., Theerthagiri, P.R., Rakeshkumar, S.: Design and analysis of genetic algorithm based controllers for non linear liquid tank system. In: 2012 International Conference on Advances in Engineering, Science and Management (ICAESM), pp. 616–620 (2012, March)
34. Bi, J., Liu, D., Zhan, K.: PID parameters optimization for liquid level control system based on genetic algorithm. JDCTA: Int. J. Digital Content Technol. Appl. **6**(1), 361–368 (2012)
35. Xiao-Gen, S., Li-Qing, X., Cheng-Chun, H.: Optimization of PID parameters based on genetic algorithm and interval algorithm. In: Control and Decision Conference, 2009. CCDC'09. Chinese, pp. 741–745 (2009, June)
36. Yuan, G., Xue, Y.G., Liu, J.: Adaptive immune genetic algorithm and its application in PID parameter optimization for main steam temperature control system. In: 2010 Third

International Workshop on Advanced Computational Intelligence (IWACI), pp. 304–309 (2010)

37. Zhang, J., Zhuang, J., Du, H.: Self-organizing genetic algorithm based tuning of PID controllers. Inf. Sci. **179**(7), 1007–1018 (2009)

38. Lin, G., Liu, G.: Tuning PID controller using adaptive genetic algorithms. In: 2010 5th International Conference on Computer Science and Education (ICCSE), pp. 519–523 (2010)

39. Rani, M.R., Selamat, H., Zamzuri, H. et al.: PID controller optimization for a rotational inverted pendulum using genetic algorithm. In: 2011 4th International Conference on Modeling, Simulation and Applied Optimization (ICMSAO), pp. 1–6 (2011)

40. Korkmaz, M., Aydogdu, Ö., Dogan, H.: Design and performance comparison of variable parameter nonlinear PID controller and genetic algorithm based PID controller. In: 2012 International Symposium on Innovations in Intelligent Systems and Applications (INISTA), pp. 1–5 (2012)

41. Ohri, J., Kumar, N., Chinda, M.: An improved genetic algorithm for PID parameter tuning. In: Proceedings of the 2014 International Conference on Circuits, Systems, Signal Processing (2014)

42. Saad, M.S., Jamaluddin, H., Darus, I.Z.: PID controller tuning using evolutionary algorithms. Wseas Trans. Syst. Control **7**(4), 139–149 (2012)

43. Jaen-Cuellar, A.Y., Romero-Troncoso, R.D.J., Morales-Velazquez, L., et al.: PID-controller tuning optimization with genetic algorithms in servo systems. Int. J. Adv. Rob. Syst. **10**, 324 (2013)

44. Sadasivan, J., Mammen, O.: Genetic algorithm based parameter identification of three phase induction motors. Reproduction **31**(10) (2011)

45. Megherbi, A.C., Megherbi, H., Benmahamed, K., et al.: Parameter Identification of Induction Motors using Variable-weighted Cost Function of Genetic Algorithms. J. Electr. Eng. Technol. **5**(4), 597–605 (2010)

46. Nithya Rani, N., Giriraj Kumar, S.M., Anantharaman, N.: Modeling and control of temperature process using genetic algorithm. Int. J. Adv. Res. Electr. Electron. Instrum. Eng. **2**(11) (2013)

47. Houck, C.R., Joines, J., Kay, M.G.: A genetic algorithm for function optimization: a Matlab implementation. NCSU-IE TR, **95**(09) (1995)

48. Guldogan, E.U., Bulut, O., Tasgetiren, M.F.: A dynamic berth allocation problem with priority considerations under stochastic nature. In: Advanced Intelligent Computing Theories and Applications. With Aspects of Artificial Intelligence, pp. 74–82. Springer, Berlin Heidelberg (2012)

49. Meffert, K., Rotstan, N., Knowles, C. et al.: Jgap-java genetic algorithms and genetic programming package. http://jgap.sf.net (2012)

50. Patrascu, M., Ion, A.: GAOT-ECM: Extension for Control and Modeling. http://www.mathworks.com/matlabcentral/fileexchange/51072-gaot-ecm–extension-for-control-and-modeling (2015). Accessed 7 Jul 2015

51. Patrascu, M., Ion, A. GAOT-ECM Seismic Vibration Case Study. http://www.mathworks.com/matlabcentral/fileexchange/51131-gaot-ecm-seismic-vibration-case-study (2015). Accessed 7 Jul 2015

52. Festo: Festo Water Tank Workstation. http://www.festo-didactic.com (2015). Accessed 7 Jul 2015

53. LD DIDACTIC Group: ELWE Technik Air Mass and Temperature System. http://www.elwe-technology.com (2015). Accessed 7 Jul 2015

54. Sims, N.D., Stanway, R., Johnson, A.R.: Vibration control using smart fluids: a state-of-the-art review. Shock Vib. Dig. **31**(3), 195–203 (1999)

55. Patrascu, M., Dumitrache, I., Patrut, P.: A comparative study for advanced seismic vibration control algorithms. UPB Sci. Bull. Series C, **74**(4):3–16 (2012)

56. Patrascu, M., Dumitrache, I. Hybrid geno-fuzzy controller for seismic vibration control. In: 2012 UKACC International Conference on Control (CONTROL), pp. 52–57 (2012, September)

57. Spencer, Jr B.E., Carlson, J., Sain, M.K. et al. (1997, June). On the current status of magnetorheological dampers: seismic protection of full-scale structures. In American Control Conference, 1997. Proceedings of the 1997, vol. 1, pp. 58–462
58. Yan, G., Zhou, L.L.: Integrated fuzzy logic and genetic algorithms for multi-objective control of structures using MR dampers. J. Sound Vib. **296**(1), 368–382 (2006)
59. Patrascu, M.: Genetically enhanced modal controller design for seismic vibration in nonlinear multi-damper configuration. Proc. Inst. Mech. Eng. Part I: J. Syst. Control Eng. **229**(2), 158–168 (2015)

Designing Fuzzy Controller for a Class of MIMO Nonlinear Systems Using Hybrid Elite Genetic Algorithm and Tabu Search

Nesrine Talbi and Khaled Belarbi

Abstract This chapter presents a Hybrid Elite Genetic Algorithm and Tabu Search (HEGATS) to design optimal fuzzy controllers for multi input multi output (MIMO) nonlinear system. The principle of the proposed algorithm is to seek the elitism by GA and introduce it in the TS algorithm as initial solution in order to find the best fuzzy rule base of the fuzzy controller. The fuzzy rule base of the fuzzy controller is tuned for optimal control performance using HEGATS by minimizing the mean square error. The proposed algorithm is tested for control of a helicopter model simulator and a double inverted pendulum. Simulation results proved the effectiveness of the proposed algorithm.

Keywords Elite genetic algorithm · Tabu search · MIMO nonlinear system · Fuzzy controller · Hybrid algorithm

1 Introduction

Most non-linear systems can be modeled under sometimes rather restrictive assumptions that can make difficult the implementation of the resulting control schemes. It is therefore necessary to consider all the imprecise and uncertain information about the system. The theory of fuzzy sets developed by Lotfi Zadeh in 1965 [1], has treated the inaccuracies and uncertainties. Many applications of fuzzy logic have been developed in various fields of modeling and control of systems, where no deterministic model exists or it is difficult to obtain.

N. Talbi (✉)
Faculty of Sciences and Technology, Department of Electronic, Jijel University, Jijel, Algeria
e-mail: t_nesrine2003@yahoo.fr

K. Belarbi
Faculty of Engineering, Mentoury University, Constantine, Algeria
e-mail: kbelarbi@yahoo.com

© Springer International Publishing Switzerland 2016
H.E. Ponce Espinosa (ed.), *Nature-Inspired Computing for Control Systems*,
Studies in Systems, Decision and Control 40, DOI 10.1007/978-3-319-26230-7_5

The first developments in fuzzy control were performed by Mamdani [2, 3]. The basic idea was to use the experience of human operators to build a control law. A set of fuzzy rules then translated the behavior of operators in terms of control strategy. Numerous studies have justified writing the rules by analogy to the sliding regime [4] or by a qualitative analysis of behavior in the phase plane [5]. These control structures are equivalent to conventional controllers [6, 7].

The advantage of a fuzzy inference system (FIS) is that only the knowledge of the process behavior is sufficient to control the synthesis of the control law. They raise a wide interest, both theoretical and practical, in the identification and control of complex and nonlinear processes. This is due to the fact that firstly the FIS does not require the existence of an analytical model of the process to control and little information is sufficient to implement the control loop. Secondly, the FIS are nonlinear systems and thus are more suitable for the control of non-linear processes.

The design of a fuzzy inference system involves the determination of its structure and its parameters. In many cases, the structure is determined empirically by choosing a priori the type of language or relational approximate reasoning, the number of fuzzy sets for each input variable, and taking all possible combinations to build the rule base. The successive attempt by adjusting the number of parameters is quite long and tedious. Various techniques of tuning, optimization and learning fuzzy systems have been developed. In 1990, neural networks were used for setting parameters of the premise membership functions [8] and fuzzy rules consequences [9–13]. The first method of tuning the parameters of premises and consequences of FIS by gradient descent was by Nomura et al. [14, 15]. A different approach of supervised learning of FIS is presented in 1993 by Jang where he proposes an adaptive neuro-fuzzy system (ANFIS) [16]. Afterwards, some improvements have been made on the neuro-fuzzy algorithms thereafter [17–19].

Other techniques of designing fuzzy control use metaheuristics. These optimization methods are heuristics to wide field of application as opposed to simple heuristics developed to solve a particular problem. Several metaheuristics are inspired by biological systems where among them there are Evolutionary Algorithms (EA) which the best known are genetic algorithms (GA) [20]. Their basic principles are the ability of populations of living organisms to adapt to their environment using selection mechanisms and genetic inheritance [21, 22]. Thrift [23] was the first to describe a method of optimization of fuzzy rules by a standard genetic algorithm, using three bits to encode each rule. At the same time, Karr [24] provides a method to intervene in the fuzzy controller with rules specified by a human expert and rules optimized by a genetic algorithm. Some developments have been made thereafter [25–28].

Later, other metaheuristics have been applied for setting the fuzzy inference systems. For example, the particle swarm optimization (PSO), developed by Kennedy and Eberhart [29] was used to optimize, in [30] and [31], a neuro-fuzzy network and, in [32], an adaptive fuzzy controller.

Genetic Algorithms (GA) have been widely used for the design of fuzzy controllers [33–37], however, they are computationally expensive because they simultaneously manage multiple solutions and they can be trapped in local minima.

To improve the quality of the solutions and to accelerate the convergence of GA, it is interesting to hybridize the GA with a local search algorithm. The role of the local search method is to explore in depth a particular area guiding the search of the genetic algorithms in this area.

Compared to the GA, Tabu Search (TS) has an interesting feature [38, 39]. It has a memory called "tabu list", which stores the right solutions, and it is characterized by its ability to avoid local solutions and avoid recycling using the flexible memory of the search history.

The main motivation of this chapter is the implementation of the capabilities offered by the hybridization of global search algorithm, Genetic Algorithm, and local search algorithm, Tabu Search, in order to develop an approach to design fuzzy Takagi Sugeno controller of type zero order for dynamic, complex and highly nonlinear systems. The hybrid algorithm adjusts the triangular membership functions of input and fuzzy singletons of output. Control problems are considered to study the performances of the proposed algorithm.

This chapter is organized as follows: Sect. 2 presents the general structure of the fuzzy controller to be optimized. In Sect. 3 the Elite GA is explained, after, in Sect. 4 Tabu Search algorithm is described. Optimization structure is detailed in Sect. 5, including the characteristics of GA and TS algorithms used in this paper. Hybrid Elite Genetic Algorithms and Tabu Search (HEGATS) is introduced in Sect. 6. Finally, the proposed algorithm is tested to generate fuzzy rule base of fuzzy controllers of two MIMO nonlinear systems: helicopter simulator and double inverted pendulum. Results of simulation are given in Sect. 7. The conclusion is given in the last section.

2 Fuzzy Controller Structure

Consider a fuzzy controller of type zero order (TSZO) with two inputs, the error $e(t)$ and its variation $\Delta e(t)$ and, an output $u(t)$, the fuzzy control applied to the system as shown in Fig. 1. Where, $e(t) = y_d(t) - y(t)$ is the difference between the desired output $y_d(t)$ and the measured output $y(t)$ of the controlled system. To have

Fig. 1 Structure of the fuzzy controller to optimize

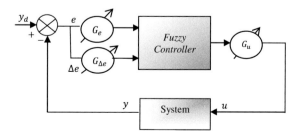

flexibility in implementing the control, the universe of discourse of the input and
output are limited to an interval $[-1, 1]$ determined by the normalization of the
input and the output. For the normalization, gain adjustments (scale factors) are
used for the desired dynamics.

For the parametric optimization of a fuzzy controller, you must define:

- The type and number of fuzzy sets for each input and output variable.
- All parameters of the membership functions that form the fuzzy partition of
  linguistic variable.
- The structure of the fuzzy rules.
- The inference mechanism, connections operators and defuzzification method.

By their structure, Takagi Sugeno (TS) fuzzy systems provide a simple analytical expression for the generated output based on inputs considered. This
property allows them to operate numerical optimization mechanisms for the synthesis of fuzzy controllers. Thus, Takagi Sugeno [40] uses the least squares algorithms while Bersini [41] uses a gradient descent method for minimizing a quadratic
criterion. Subsequently, numerous studies have been published on the use of TS
type fuzzy controllers in the control of nonlinear processes.

2.1 Fuzzification

In this chapter, the membership function of the adopted structure has only three
fuzzy sets for each input, equidistant triangular membership functions in a normalized universe of discourse on $[-1, 1]$, as shown in Fig. 2, and equidistant
singleton membership functions for the output. The inputs $e(t)$ and $\Delta e(t)$ are
fuzzified by following membership functions:

$$\mu_N(x) = \begin{cases} 1 & \text{if } x < a_1 \\ \frac{a_2 - x}{a_2 - a_1} & \text{if } a_1 \leq x < a_2 \\ 0 & \text{if } x \geq a_2 \end{cases}$$

$$\mu_Z(x) = \begin{cases} \frac{x - a_1}{a_2 - a_1} & a_1 \leq x < a_2 \\ \frac{a_3 - x}{a_3 - a_2} & a_2 \leq x < a_3 \\ 0 & \text{otherwise} \end{cases} \qquad (1)$$

$$\mu_P(x) = \begin{cases} 0 & x < a_2 \\ \frac{x - a_2}{a_3 - a_2} & a_2 \leq x < a_3 \\ 1 & x \geq a_3 \end{cases}$$

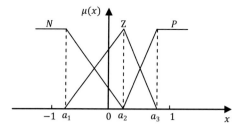

Fig. 2 Partition of fuzzy controller inputs

where x is the variable to the fuzzifier. N, P and Z are fuzzy sets of the input variable; their meaning is successively Negative, Positive and Zero. Since the fuzzy controller is of type TSZO, fuzzy rules conclusions are constants (fuzzy singletons).

2.2 Fuzzy Rule Base

Based on the description of the system to control, linguistic variables and membership functions for input and output variables, we can establish the inference rules. The ith rule of a fuzzy controller type TSZO is presented as follows:

$$R_i : \text{ If } e\,(k) \text{ is } A_i \text{ and } \Delta e(k) \text{ is } B_i \text{ then } u\,(k) \text{ is } C_i \tag{2}$$

where R_i is the ith rule, $x_i(k)$ is the input variable, k is the time step and $u(k)$ is the control output variable. A_i and B_i are fuzzy linguistic value and C_i is a fuzzy singleton [40].

The number of fuzzy rules decision is a very important issue as it plays a key role in the control and modeling of fuzzy systems. There is generally no systematic and effective procedure for selecting the most appropriate number of rules. A reasonable number of fuzzy rules, without losing too much information about the system to be controlled or model must be carefully obtained.

In this chapter, only three rules have been proposed:

- IF e is $(-1, a_1, a_2)$ and Δe is $(-1, b_1, b_2)$ then u is C_1.
- IF e is (a_1, a_2, a_3) and Δe is (b_1, b_2, b_3) then u is C_2.
- IF e is $(a_2, a_3, 1)$ and Δe is $(b_2, b_3, 1)$ then u is C_3.

2.3 Inference Engine and Defuzzification

In the inference engine, the fuzzy AND operation is implemented by the algebraic product in fuzzy logic theory. Thus, given an input data set $\vec{x} = (e, \Delta e)$, the firing strength $\gamma_i(\vec{x})$ of rule i is calculated by

$$\gamma_i\left(\vec{x}\right) = \prod_{j=1}^{2} \mu_{ij}(x_j) \qquad (3)$$

If there are r rules in an FLC, the output of the controller calculated by the weighted average defuzzification method is

$$u = \frac{\sum_{i=1}^{r} \gamma_i\left(\vec{x}\right) C_i}{\sum_{i=1}^{r} \gamma_i\left(\vec{x}\right)} \qquad (4)$$

where C_i is the rule consequent value in (1).

The optimization of an FLC consists of the determination of all free parameters in each fuzzy rule.

3   Elite Genetic Algorithm

3.1   The Biological Metaphor for GA

The biological metaphor for genetic algorithms is the evolution of the species by survival of the fittest, as described by Charles Darwin. In a population of animals or plants, a new individual is generated by the crossover of the genetic information of two parents. The genetic information for the construction of the individual is stored in the DNA. The human DNA genome consists of 46 chromosomes, which are strings of four different bases, abbreviated A, T, G and C. A triple of bases is translated into one of 20 amino acids or a "start protein building" or "stop protein building" signal [42]. In total, there are about three billion nucleotides. These can be structured in genes, which carry one or more pieces information about the construction of the individual. However, it is estimated that only 3 % of the genes carry meaningful information, the vast majority of genes—the "junk" genes—is not used [43].

The genetic information itself, the genome, is called the genotype of the individual. The result, the individual, is called phenotype. The same genotype may result in different phenotypes.

A genetic algorithm tries to simulate the natural evolution process. Its purpose is to optimize a set of parameters. It is originally proposed by Holland in 1975 [44], as a search algorithm. It succeeded in optimizing complicated functions, especially nonlinear problems, which are difficult to solve by other search methods. Genetic Algorithms, as the name implies, simulate the mechanisms of reproduction of living organisms to find an optimal solution to a given problem. The search for solutions is through operations such as reproduction, crossing and mutation.

3.2 EGA Characteristics

Before getting in a GA details, it is necessary to explain some terms:

- *Chromosome (individual)*: a set of genes; a chromosome contains the solution in the form of genes.
- *Gene*: a part of a chromosome; a gene contains a part of a solution; it determines the solution.
- *Population*: number of individuals present with same chromosome length.
- *Fitness*: the value assigned to an individual based on how far or close a individual is from the optimal solution.
- *Reproduction*: a random mix of chromosomes of two individuals, giving birth to children individuals with a new genetic fingerprint, inherited from parents.
- *Mutation*: changing a random gene in an individual.
- *Selection*: selecting individuals for creating the next generation.

The elitist strategy is to maintain the best individual of each generation. Elitism prevents the best individual from disappearing in the selection, or that good combinations are affected by breeding and mutation operators. After each evaluation of the performance of the individuals in a generation t, the best individuals of the preceding generation $(t - 1)$ are reintroduced in the population if none of the people of the generation t are better than him. By this approach, the performance of the best individual of the current population is monotonically from generation to generation. The pseudo-code of the EGA is as follows:

```
BEGIN

   Initialize population with random candidate solution
   Evaluate each candidate
   Repeat until (termination condition) is satisfied

      Select parents;
      Recombine pairs of parents;
      Mutate the resulting offspring;
      Select individuals of the next generation and add
      elitism to the new generation.

   END

END
```

4   Tabu Search

Tabu search (TS) is an iterative procedure designed for the solution of optimization problems. TS was invented by Glover in 1986 and has been used to solve a wide range of hard optimization problems such as job-shop scheduling, graph coloring problems, the 'traveling salesman problem' (TSP) and the capacitated arc routing problem. TS can be considered as a 'meta-strategy' for guiding known heuristics to overcome local optimality [38, 39].

TS is based on the premise that problem solving, in order to be qualified as intelligent, must incorporate adaptive memory and responsive exploration. The adaptive memory feature of TS allows the implementation of procedures that are capable of searching the solution space economically and effectively. The role of TS will most often be to guide and orient the search of another search procedure [45, 46]. To avoid cycling, solutions that were recently examined are declared 'tabu' for a certain number of iterations. Applying intensification procedures can accentuate the search in a promising region of the solution space. In contrast, diversification can be used to broaden the search to a less explored region [39]. Although still in its infancy stages, this metaheuristic has been reported in literature during the last few years as successful solution approaches for a great variety of problems [47].

The TS algorithm can be summarized as follows:

```
Randomly create an initial solution s₀, let current
solution s := s₀.

Let best solution s *:= s ; f * = f (s)

tabu list T: = {}

Repeat

-    m := the best move among the not tabu movements
     and the exceptional tabu movements.
-    s : = s ⊕ m
-    IF f (s) < f ( s *) DO
          s *: =s ; f * := f (s)
-    Update T

Until < stopping criteria satisfied >

Return s*
```

The stopping criteria condition can be defined when the best solutions are found or when the time limits have been reached.

5 Optimization Method

5.1 Optimization Structure

The diagram of the closed loop control is shown schematically in Fig. 3. It consists of four main blocks:

- Structural block represented by the proposed fuzzy controller [FLC].
- Block of the system to control [Plant].
- Optimization block characterized by a hybrid algorithm [Hybrid Algorithm].
- Decision block characterized by the desired performance criteria [Performance Index].

The mutual interaction between the different blocks of the structure of Fig. 3 is illustrated by the following procedure:

1. Generation of an initial population of chromosomes characterizing the controller settings.
2. For all chromosomes:

   - Evaluate the objective function.
   - Classify chromosomes according to their ability.
   - Construction of a new population by applying the genetic operators adapted to the encoding of chromosomes.

Step 2 is repeated until a maximum number of generations is performed. After the process of evolution, the final generation of the algorithm consists of the well-adjusted individuals and that provide optimal or close solutions.

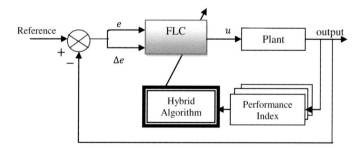

Fig. 3 Optimization and control structure

The inclusion of design constraints in the optimization process preserves the semantics of the fuzzy rules. For this, the constraints on the limits of the parameters vector to be identified, and limits on control variables are imposed.

5.2 Chromosome Structure

To apply the GA directly or coupled with other meta-heuristics, the problem of the representation of the chromosome is posed. The first decision that the designer must make is how to represent a solution in a chromosome structure.

The chromosome used in this chapter has nine parameters; these parameters represent the starting point, vertex point and end point for a triangle membership function belonging to the inputs and fuzzy singleton for its output, while respecting the following constraints:

$$\begin{cases} a_1 < a_2 < a_3 \\ b_1 < b_2 < b_3 \\ C_1 < C_2 < C_3 \end{cases}$$

The chromosome is defined by vector of parameters P as:

$$P = [a_1 \, a_2 \, a_3 \, b_1 \, b_2 \, b_3 \, C_1 \, C_2 \, C_3]$$

5.3 Main Algorithm of GA

In order to build an optimization algorithm based on GA, we must prepare the following platform:

- Define the initial configuration (size of the population, stopping criterion, crossover and mutation probabilities, etc.).
- Define the appropriate mode of encoding (binary, Real, etc.).
- Define the generation method of the initial population.
- Define the objective function.
- Adapt genetic operators to the mode of selected coding.

The GA used for the study in this chapter has the following basic characteristics:

- The encoding used for the chromosome is Real numbers, since the resolution of the problem of optimization will be more effective with GA real encoding.
- The initial population is generated randomly to cover all of the search space.

- The method of "biased lottery" is chosen for the selection operation, so that the individuals who have a great fitness are more luckily to be selected.
- Crossing in 2 points is used to make it possible to introduce more of diversity. Indeed, more the number of points of crossings is large, more the probability of crossing will be more high, there will be an exchange of segments, therefore an exchange of parameters and information.
- The uniform mutation is chosen because it is the simplest and most suitable for the real encoding.
- The final selection for the new generation will be a competition where the best individuals (children and parents) are involved in the new generation.

The overall objective of the control system is to minimize the error each time k between the actual system response $y(k)$, and the reference $y_d(k)$. There are several performance indicators to measure the quality of a given setting, the performance criteria used in this chapter is the Mean Square Error (MSE) defined by:

$$F = 1/nT \sum_{k=1}^{n} e(k)^2 \qquad (5)$$

where n is the total number of samples, and T the sampling time, and $e(k) = r(k) - y(k)$ is the difference between the value of the desired output $r(k)$, and the value of the measured output $y(k)$ of the process under control.

Genetic algorithms are used for determining the maximum of a function, so the fitness function of the GA is defined as:

$$f = 1/(1+F) \qquad (6)$$

5.4 Features of the TS Algorithm

Before building an optimization algorithm Tabu Search (TS), it is necessary to define:

- The initial configuration (size of the Tabu list, number of neighboring solutions, stopping criteria).
- The objective function.
- The method of generating the initial solution and its vicinity.

The basic features of the used TS algorithm in this study are:

- Generation of a random neighboring solutions and initial solution, taking into account all the search space and respecting the intensification and diversification criteria.

- The tabu list is a FIFO memory (First Input First Output); its size is often proportional to the number of iterations.
- The fitness function will automatically follow the choice of the performance criterion of the GA calculated by formula (5).

6 HEGATS Optimization Algorithm

A traditional GA often explores the candidate solution encoded in chromosomes and exploits those with better fitness iteratively until the solution is reached. The disadvantages of a GA are, its ability to "climb" and the poor local research capacity, which usually lead to premature convergence and local optimum. Experiments show that when research has reached a certain level, a GA is often much inactive redundant iteration, which reduces the efficiency [48, 49]. Even, it appears that elitism significantly improves the performance of the genetic algorithm for certain classes of problems, but it can degrade for other classes, by increasing the rate of premature convergence. Therefore, it may be advantageous to integrate various techniques of search, such as TS, to form a hybrid method which guarantees the global convergence of the specific problem. In this study we propose to combine the GA with Elitism with Tabu Search to accelerate the convergence of the GA and find a better solution in a minimum computation time and high accuracy.

At each iteration of the GA, the best chromosome of the actual generation, $Elit$, is introduced in the TS algorithm to improve its quality by searching in its neighborhood another best solution $Elit_{best}$; this latter will be included in the new population of the next generation of the GA. Table 1 defines the acronyms used in the HEGATS optimization algorithm.

The HEGATS learning algorithm can formulated as follows:

Table 1 Parameters of the HEGATS algorithm

Parameter	Significance
N	Population size
Pc	Crossing probability
Pm	Mutation probability
$Elit$	Elitist solution
$Elit\_Best$	Best elitist solution
s_init	Initial solution introduced in TS algorithm
k	Maximum number of iterations in TS algorithm
Nv	Number of neighboring solutions in TS algorithm
TL	Size of the tabu list
max_gen	Maximum generations of GA
s_opt	Optimal solution

```
Begin Genetic Algorithm
Set max_gen, Pc, Pm, N
Generate the initial population

    For i = 1 to max_gen Do

        Evaluate each chromosome in the population by eq.6
        Select the parents of the population
        Producing  the  children's  parents  selected  by  the
        crossover operation
        Mutate chromosomes of the resulting population
        Select  children  and  the  most  suitable  parents  for
        the  next  generation  and  record  the  best  chromosome
        as Elit

        Begin Tabu Search

        Set  k,  Nv,  T,  Initialize  tabu  list  T = ∅,  s_init  =
        Elit

        While i < k

            Generate Nv solutions neighboring of s_init
            Evaluate each neighbor solution by eq.5
            Take the most appropriate solution as Elit_best
            Update tabu list
            i = i + 1

        End While

        End Tabu Search

        Save the best individual as Elit_best and insert it
        in the new generation

    End For

Return the best solution s_opt

End
```

7 Simulation and Interpretation

To demonstrate the efficiency of the proposed method, HEGATS is tested to generate automatically fuzzy rule base for control of two MIMO nonlinear systems: a helicopter simulator and a double inverted pendulum.

7.1 Stabilization of a Helicopter Simulator

7.1.1 Helicopter Simulator Model

The CE150 Helicopter model of Humusoft Ltd consists of a body carrying two DC motors. These motors drive the propellers. The body has two degrees of freedom. The axes of the body rotation are perpendicular as well as the axes of the motors (see Fig. 4). Both body position angles, i.e. azimuth angle in horizontal (Φ) and elevation angle in vertical plane (Ψ) are influenced by the rotating propellers simultaneously [50, 51].

The helicopter model is a multivariable dynamical system with two manipulated inputs u_1 and u_2 and two measured outputs Φ and Ψ. All inputs and outputs are coupled. The system is essentially nonlinear and at least of the sixth order, depending on the modeling precision. The user of the simulator communicates with the system via the data-processing interface [52].

The mathematical model of the helicopter simulator is given by the following differential equations [49]

Fig. 4 Helicopter model

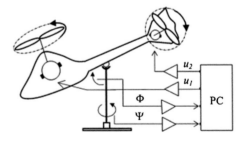

$$\dot{x}_1 = x_2$$

$$\dot{x}_2 = 0.8764x_2 sinx_1 + 3.4325x_4 u_1 cosx_1 + 0.4211x_2 - 0.0035x_5^2$$
$$\quad - 46.35x_6^2 - 0.8076x_5x_6 - 0.0259x_5 - 2.9749x_6$$

$$\dot{x}_3 = x_4$$

$$\dot{x}_4 = 21.4010x_4 - 31.8841x_8^2 - 14.2029x_8 - 21.7150x_9 + 1.4010u_1$$

$$\dot{x}_5 = -6.6667x_5 - 2.7778x_6 + 2u_1$$

$$\dot{x}_6 = 4x_5$$

$$\dot{x}_7 = -8x_7 - 4x_8 + 2u_2$$

$$\dot{x}_8 = 4x_7$$

$$\dot{x}_9 = -1.3333x_9 + 0.0625u_1$$

(7)

where $x_1 = \Psi, x_2 = \frac{d\Psi}{dt}, x_3 = \Phi, x_4 = \frac{d\Phi}{dt}$, and x_5 to x_9 are state variables representing the two DC motors and the coupling effects.

7.1.2 Control Structure

The objective is to stabilize the helicopter around a set point (Ψ_r, Φ_r). For that, decentralized control method is used; elevation and azimuth subsystems are controlled by two fuzzy controllers (see Fig. 5).

In this study, each controller is characterized by three triangular membership functions for inputs and three singletons for output (see Fig. 2), and the rule base described in Sect. 2.2. The tuning parameters vector (chromosome) is composed by the parameters of the two fuzzy controllers (see Sect. 5.2), therefore 18 parameters. The GA and TS characteristics are described in Sects. 5.3 and 5.4 successively.

Fig. 5 Control structure of helicopter simulator [53]

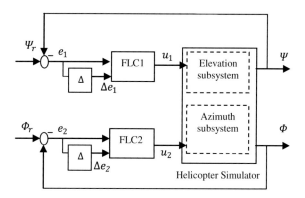

7.1.3 Simulations Results

The overall objective of the control system is to minimize the error at each time k between the actual response of the system and the set point, the performance index (Mean Square Error), F, is chosen as follows:

$$F = \frac{1}{nT} \left(\sum_{k=1} e_1^2(k) + \sum_{k=1} e_2^2(k) \right) \tag{8}$$

where e_1 is elevation angle error, e_2 is azimuth angle error, n is the total number of samples, and T the sampling time. In the simulation, we use the following parameters: $T = 0.1, \text{max\_gen} = 100, N = 10, Pc = 0.9, Pm = 0.07, TL = 10, Nv = 8, k = 4$.

The HEGATS algorithm was run for 100 iterations. After each iteration of EGA, TS was run for 4 iterations to find the best solution, the last iteration of the HEGATS learning algorithm contained the optimal parameters of the two controllers.

The learned fuzzy rule base of FLC1 is as follows:

Rule 1 : if e_1 is $(-0.1, -0.09, 0.003)$ and Δe_1 is $(-0.05, -0.02, 0.01)$ then $u_1 = -0.29$
Rule 2 : if e_1 is $(-0.09, 0.003, 0.06)$ and Δe_1 is $(-0.02, 0.01, 0.02)$ then $u_1 = -0.17$
Rule 3 : if e_1 is $(0.003, 0.06, 0.1)$ and Δe_1 is $(0.01, 0.02, 0.05)$ then $u_1 = 0.3$

The learned fuzzy rule base of FLC2 is as follows:

Rule 1 : if e_2 is $(-0.4, -0.15, -0.03)$ and Δe_2 is $(-0.2, -0.14, -0.03)$ then $u_2 = -0.07$
Rule 2 : if e_2 is $(-0.15, -0.03, 0.13)$ and Δe_2 is $(-0.14, -0.03, 0.12)$ then $u_2 = -0.009$
Rule 3 : if e_2 is $(-0.03, 0.13, 0.4)$ and Δe_2 is $(-0.030, 0.12, 0.2)$ then $u_2 = 0.08$

Figure 6 shows the evolution of the fitness function. In Fig. 7, the elevation and azimuth angles using optimal controllers are shown. The Fig. 8 shows the control signal u_1 for the elevation subsystem and the control signal for the azimuth subsystem using the learned parameters values obtained by HEGATS learning algorithm. It is clear that the proposed algorithm can stabilize the helicopter simulator in less time (and less than 20 iterations) with best accuracy equal to 0.1123.

7.2 Stabilization of Double Inverted Pendulum

The structure of the double inverted pendulum system is illustrated in Fig. 9. The double inverted pendulum is connected by a return spring, two couples u_1 and u_2 are applied to the actuators through inputs to ensure the perpendicular position.

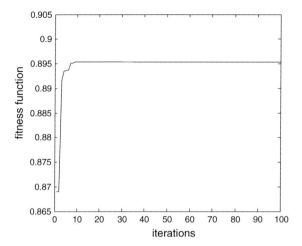

Fig. 6 Evolution of the fitness function

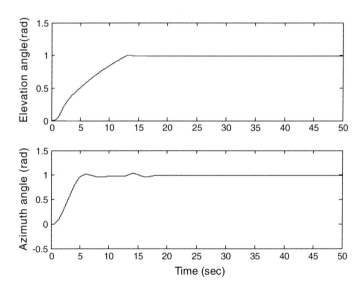

Fig. 7 Response of elevation and Azimuth angles

The dynamics of the system [54] are represented by:

$$\begin{cases} \dot{x}_1 = x_2 \\ \dot{x}_2 = \theta_1 sinx_1 + \theta_2 sinx_3 + \theta_3 + \frac{u_1}{J_1} \\ \dot{x}_3 = x_4 \\ \dot{x}_4 = \theta_4 sinx_3 + \theta_5 sinx_1 + \theta_6 + \frac{u_2}{J_2} \end{cases} \quad (9)$$

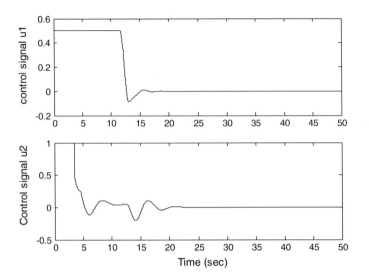

Fig. 8 Control signals u_1 and u_2

Fig. 9 Structure of the
double inverted pendulum

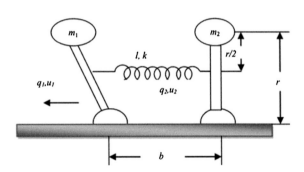

where

$$\theta_1 = \frac{m_1 g r}{J_1} - \frac{k r^2}{4 J_1}, \theta_2 = \frac{k r^2}{4 J_1}, \theta_3 = \frac{k r}{2 J_1}(l - b),$$

$$\theta_4 = \frac{m_2 g r}{J_2} - \frac{k r^2}{4 J_2}, \theta_5 = \frac{k r^2}{4 J_2}, \theta_6 = \frac{k r}{2 J_2}(l - b).$$

The state variables are the angular positions and angular velocities:

$$\begin{bmatrix} x_1 \\ x_2 \\ x_3 \\ x_4 \end{bmatrix} = \begin{bmatrix} q_1 \\ \dot{q}_1 \\ q_2 \\ \dot{q}_2 \end{bmatrix},$$

Table 2 Parameters of the HEGATS algorithm for double inverted pendulum

Parameter	Value
N	10
Pc	0.8
Pm	0.07
k	4
Nv	8
TL	7
max_gen	10

where $m_1 = 2$ kg, $m_2 = 2.5$ kg, $k = 100$ N/m, $l = 0.5$ m, $r = 0.5$ m, $b = 0.4$ m, $J_1 = 0.5$ and $J_2 = 0.625$.

The objective is to define two couples to maintain the equilibrium state defined by $q_1 = 0$ rad and $q_2 = 0$ rad. The initial conditions are $(q_1(0), q_2(0)) = (0.5$ rad, -0.5 rad).

The specific values of the hybrid optimization algorithm for fuzzy controllers of double inverted pendulum and the constraints are given in Table 2.

To illustrate the performance of the proposed method, we consider the control of the double inverted pendulum system whose dynamics are given by the expression (9). The control structure comprises two fuzzy controllers working separately: a fuzzy controller for angle q_1 and the other one for angle q_2.

Figure 10 shows the evolution of the fitness function and performance status during the execution of the HEGATS algorithm.

The fuzzy rule base of FLC1 obtained after optimization is as follows:

Rule 1 : *if* e_1 *is* N $(-1, -0.96, 0.43)$ *and* Δe_1 *is* N $(-1, -0.83, -0.23)$ *then* $u_1 = -0.56$
Rule 2 : *if* e_1 *is* Z $(-0.96, 0.43, 0.96)$ *and* Δe_1 *is* Z $(-0.83, -0.23, 0.53)$ *then* $u_1 = -0.64$
Rule 3 : *if* e_1 *is* P $(0.43, 0.96, 1)$ *and* Δe_1 *is* $P(-0.23, 0.53, 1)$ *then* $u_1 = 0.96$.

Fig. 10 Evolution of fitness function

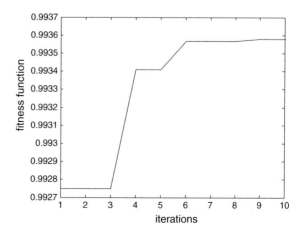

The fuzzy rule base of FLC2 obtained after optimization is as follows:

Rule 1 : *if e_2 is N $(-1, -0.83, 0.43)$ and Δe_2 is N $(-1, -0.80, -0.50)$ then $u_2 = -0.63$*
Rule 2 : *if e_2 is Z $(-0.83, 0.43, 0.83)$ and Δe_2 is Z $(-0.80, -0.50, 0.93)$ then $u_2 = 0.16$*
Rule 3 : *if e_2 is P $(0.43, 0.83, 1)$ and Δe_2 is P $(-0.50, 0.93, 1)$ then $u_2 = 0.80$.*

Figure 11 shows the responses of angular positions q_1 and q_2, the angular velocities $\dot{q}_1$ and $\dot{q}_2$ and the two couples of command u_1 and u_2. The results show the convergence of the angles and angular velocities to the equilibrium state.

The robustness of the optimized controller is justified by its ability to respond well to the change of initial operating conditions and the change of parameters of the double pendulum.

For the first case, the initial conditions are tested as follows: $(q_1(0), q_2(0)) \in \{\pm 0.3, \pm 0.5, \pm 0.8, \pm 1.2\}$ rad, where $(\dot{q}_1(0), \dot{q}_2(0)) = (0\,\text{rad/s}, 0\,\text{rad/s})$. Figure 12 shows these different situations.

The second case, robustness via the parameter change, is illustrated by an increase in the parameter $\delta = kr(l - b)$. Figure 13 shows the evolution of angular positions for 10 δ. There has been a convergence for the equilibrium with low sensitivity via this variation.

According to Figs. 11, 12 and 13, it is observable that the optimized fuzzy controllers are capable to well control the double inverted pendulum.

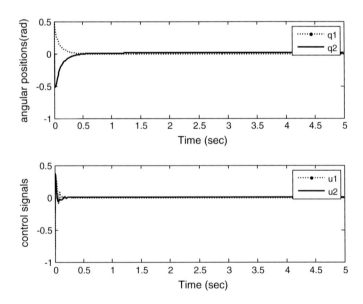

Fig. 11 System responses submitted to the optimized fuzzy controllers

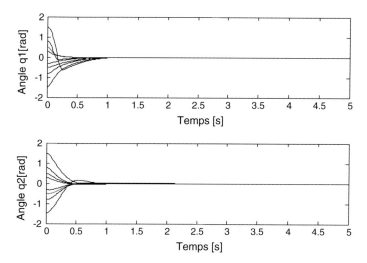

Fig. 12 System responses with different initial conditions

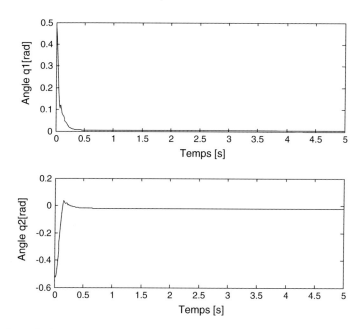

Fig. 13 Test of robustness through the change of the parameter δ

8 Conclusion

In this chapter, hybrid Elite Genetic Algorithm and Tabu Search (HEGATS) learning algorithm is used to simultaneously optimize the premise and consequent parameters of the fuzzy rules for the design of fuzzy controller of Takagi-Sugeno zero-order type. The fuzzy controllers are used to stabilize a helicopter simulator and double inverted pendulum, which are MIMO nonlinear systems. Decentralized command of two fuzzy controllers is used to calculate control systems. Simulation results have shown that the application of the proposed HEGATS algorithm for MIMO systems can design a fuzzy control system able to improve the calculation time and the response time of the process and optimize the accuracy despite the complexity of the controlled system.

References

1. Zadeh, L.A.: Fuzzy sets. Inf. Control **8**, 33 (1965)
2. Mamdani, E.: Application of fuzzy logic to approximate reasoning using linguistic systems. Fuzzy Sets Syst. **26**, 1182–1191 (1977)
3. Procyk, T.J., Mamdani, E.H.: A linguistic self-organising process controller. Automatica **15**, 15–30 (1979)
4. Palm, R.: Sliding mode fuzzy control. In: Proceeding of the IEEE Presented at the Conference on Fuzzy Systems (Fuzz´IEEE 92), pp. 519–526. San Diego, USA (1992)
5. Foulloy, L.: Contrôle qualitative et contrôle flou: vers une méthode d'écriture des contrôleurs flous. Actes du 12ième journée internationale sur les systèmes experts et leurs applications, Avignon, France (1992)
6. Galichet, S., Dussud, M., Foulloy, L.: Contrôleurs flous: equivalences et études comparatives. Actes des rencontres francophones sur la logique floue et ses applications (LFA'92), pp. 229–236. Nîmes, France (1992)
7. Galichet, S., Foulloy, L.: Fuzzy controllers: synthesis and equivalences. IEEE Trans. Fuzzy Syst. **3**(2), 140–148 (1995)
8. Hayashi, I., Nomura, H., Wakami, N.: Acquisition of inference rules by neural network driven fuzzy reasoning. Jpn. J. Fuzzy Theory Syst. **2**(4), 453–469 (1990)
9. Wang, L.X.: Fuzzy systems as nonlinear dynamic system identifiers part I: design. In: Proceedings of the 31st Conference on Decision and Control, pp. 897–902. Tucson, AZ (1992)
10. Wang, L.X., Mendel, J.M.: Back-propagation fuzzy system as nonlinear dynamic system identifiers. In: Proceedings of the First IEEE Conference on Fuzzy Systems, pp. 1409–1418 (1992)
11. Horikawa, S., Furuhashi, T., Uchikawa, Y.: Composition methods and learning algorithms of fuzzy neural networks. Jpn. J. Fuzzy Theory Syst. **4**(5), 529–556 (1992)
12. Wang, L.X.: Stable adaptive fuzzy control of nonlinear systems. IEEE Trans. Fuzzy Syst. **1**(2), 146–155 (1993)
13. Wang, L.X.: Design and analysis of fuzzy identifiers of nonlinear dynamic systems. IEEE Trans. Autom. Control **40**(1), 11–23 (1995)
14. Nomura, H., Hayashi, I., Wakami, N.: A learning method of fuzzy inference rules by descent method. In: Proceedings of the First IEEE Conference on Fuzzy Systems, pp. 203–210 (1992)
15. Nomura, H., Hayashi, I., Wakami, N.: Self-Tuning fuzzy reasoning by delta rule and its application to obstacle avoidance. Jpn. J. Fuzzy Theory Syst. **4**(2), 261–272 (1992)

16. Jang, J.S.R.: ANFIS: adaptive-network-based fuzzy inference systems. IEEE Trans. Syst. Man Cybern. **23**(3), 665–685 (1993)
17. Shi, Y., Mizumoto, M.: An improvement of neuro-fuzzy learning algorithm for tuning fuzzy rules. Fuzzy Sets Syst. **118**, 339–350 (2001)
18. Rojas, I., Piomares, H., Ortega, J., Prieto, A.: Self-organized fuzzy system generation from training examples. IEEE Trans. Fuzzy Syst. **1**, 23–36 (2000)
19. Kasabov, N.K., Song, Q.: DENFIS: dynamic evolving neural-fuzzy inference system and its application for time-series prediction. IEEE Trans. Fuzzy Syst. **10**(2), 144–154 (2002)
20. Yao, X.: Evolutionary Computation—Theory and Applications. World Scientific, Singapore (1999)
21. Goldberg, D.E., Holland, J.H.: Genetic algorithms and machine learning. Mach. Learn. **3**, 95–99 (1988)
22. Holland, J.H.: Adaptation in Natural and Artificial Systems, 2nd edn. MIT Press, Cambridge (1992)
23. Thrift, P.: Fuzzy logic synthesis with genetic algorithms. In: Proceedings of the Fourth International Conference on Genetic Algorithms, pp. 509–513 (1991)
24. Karr, C.L.: Applying genetics to fuzzy logic. AI Expert **6**(3), 38–43 (1991)
25. Mohammadian, M., Stonier, R.: Generating fuzzy rules by genetic algorithms. In: Proceeding of 3rd IEEE International Workshop on Robot and Human Communication, pp. 362–367. Nagoya (1994)
26. Herrera, F., Lozano, M., Verdegay, J.L.: Tuning fuzzy logic controllers by genetic algorithms. Int. J. Approximate Reasoning **1**, 299–315 (1995)
27. Chen, C., Wong, C.: Self-generating rule-mapping fuzzy controller design using a genetic algorithm. IEE Proceedings Control Theory Appl. **49**, 143–148 (2002)
28. Belarbi, K., Titeli, F., Bourebia, W., Benmohammed, K.: Design of Mamdani fuzzy logic controllers with rule base minimisation using genetic algorithm. Eng. Appl. Artif. Intell. **18**, 875–880 (2005)
29. Kennedy, J., Eberhart, R.C.: Particle swarm optimization. In: IEEE International Conference on Neural Networks, Piscataway, pp. 1942–1948 (1995)
30. Juang, C.F.: A hybrid of genetic algorithm and particle swarm optimization for recurrent network design. IEEE Trans. Syst. Man Cybern. **34**(2), 997–1006 (2004)
31. Chatterjee, A., Pulasinghe, K., Watanabe, K., Izumi, K.: A particle swarm-optimized fuzzy-neural network for voice-controlled robot systems. IEEE Trans. Ind. Electron. **52**(6), 1478–1489 (2005)
32. Sharma, K.D., Chatterjee, A., Rakshit, A.: A hybrid approach for design of stable adaptive fuzzy controllers employing Lyapunov theory and particle swarm optimization. IEEE Trans. Fuzzy Syst. **17**(2), 329–342 (2009)
33. Herrera, F., Lozano, M., Verdegay, J.L.: A learning process for fuzzy control rules using genetic algorithms. Fuzzy Sets Syst. **100**(1–3), 143–158 (1998)
34. Chang, C.H., Wu, Y.C.: Genetic algorithm based tuning method for symmetric membership functions of fuzzy logic control system. In: Proceedings IEEE/IAS International Conference on Industrial Automation and Control Emerging Technologies, pp. 421–428 (1995)
35. Homaifar, A., McCormick, E.: Simultaneous design of membership functions and rule sets for fuzzy controllers using genetic algorithms. IEEE Trans. Fuzzy Syst. **3**(2), 129–139 (1995)
36. Belarbi, K., Titel, F.: Genetic algorithm for the design of a class of fuzzy controllers: an alternative approach. IEEE Trans. Fuzzy Syst. **8**(4), 398–405 (2000)
37. Hoffmann, F.: Evolutionary algorithms for fuzzy control system design. Proc. IEEE **89**(9), 1318–1333 (2001)
38. Glover, F.: Tabu search—part I. ORSA J. Comput. **1**(3), 190–206 (1989)
39. Glover, F.: Tabu search—part II. ORSA J. Comput. **2**(1), 4–32 (1990)
40. Takagi, T., Sugeno, M.: Fuzzy identification of systems and its applications to modeling and control. IEEE Trans. Syst. Man Cybern. **1**(1), 116–132 (1985)

41. Bersini, H., Gorrini, V.: Methods for adaptive process control. In: Proceedings of the 1st European Congress on Fuzzy Intelligent Technologies EUFIT'93, pp. 55–61. Aachen, Germany (1993)
42. Boers, E., Kuiper, H.: Biological metaphors and the design of modular artificial neural networks. Master thesis at Leiden University, The Netherlands (1992)
43. Koenn, P.: Combining genetic algorithms and neural networks: the encoding problem. A Thesis for the Master of Science Degree, The University of Tennessee, Knoxville (1994)
44. Holland, J.: Adaption in Natural and Artificial Systems. University of Michigan Press, Ann Harbor (1975)
45. Hertz, A., Taillard, E., Werra, D.: A tutorial on tabu search. In: Proceedings of Giornate di Lavoro AIRO, vol. 95, pp. 13–24 (1995)
46. Aarts, E., Lenstra, J.K.: Local Search in Combinatorial Optimization. Wiley, Old technical report ORWP9218, p. 121–136 (1997)
47. Bagis, A.: Fuzzy rule base design using tabu search algorithm for nonlinear system modeling. ISA Trans. **47**, 32–44 (2008)
48. Garcia-Martinez, C., Lozano, M., Herrera, F., Molina, D.: Global and local real-coded genetic algorithms based on parent-centric crossover operators. Eur. J. Oper. Res. **3**(185), 1088–1113 (2008)
49. Ding, J.L., Chen, Z.Q., Yuan, Z.Z.: On the combination of genetic algorithm and Ant algorithm. J. Comput. Res. Dev. **40**(9), 1351–1356 (2003)
50. Boubertakh, H., Labiod, S., and Tadjine,M.: PSO to Design Decentralized Fuzzy PI Controllers: Application for a Helicopter. 20th Mediterranean Conference on Control & Automation (MED). Barcelona, Spain (2012)
51. Wan, E.A., Bogdanov, A.A.: Model predictive neural control with applications to a 6 DOF helicopter model. In: Proceedings 2001 American Control Conference, pp. 488–493. Arlington, Virginia (2001)
52. Boubertakh, H., Tadjine, M.: Tuning Fuzzy PD and PI controllers using reinforcement learning. ISA Trans. **49**, 543–551 (2010)
53. Talbi, N., Belarbi, K.: Designing fuzzy controllers for a class of MIMO systems using Hybrid Particle Swarm optimization and Tabu Search. Int. J. Hybrid Intell. Syst. **10**(1), 1–9 (2013)
54. Spooner, J.T., Passino, K.M.: Adaptive control of a class of decentralized nonlinear systems. IEEE Trans. Autom. Control **41**, 280–284 (1996)

Neural Network Fitting for Input-Output Manifolds in Constrained Linear Systems

José M. Araújo and Carlos E.T. Dórea

Abstract This work presents a recent contribution regarding the application of multi-layer perceptron neural networks (MLP-NNs) to the fitting of complex piecewise affine control and observation laws in constrained linear systems. Such input-output maps arise from the imposition of the optimal contraction rate trajectory for the system state, or error in the observer case, within a given invariant polyhedral set that enforces the systems constraints, or in the case of optimal control laws in model predictive control (MPC). Although an offline law can be computed via multi-parametric optimization in the state feedback case, the number of regions that define the piecewise affine law can be very large, resulting in high hardware storage requirements and difficulty in quickly locating the state. On the other hand, online laws are usually more expensive in the runtime sense and can become infeasible for application in fast dynamics systems, unless an advanced, expensive processor is employed. From the data obtained by the simulation of online laws, MLPs can be trained to emulate such maps; thus, they can replace online computation and drastically reduce the runtime. Two numerical examples, one of which is based on a two-tank hydraulic system model, are presented to illustrate the proposed approach, with detailed design for constrained error estimation as well as static and dynamic output feedback.

J.M. Araújo (✉)
Grupo de Pesquisa em Sinais e Sistemas, Instituto Federal da Bahia,
Rua Emidio dos Santos, S/N, Barbalho, 40301-015 Salvador, BA, Brazil
e-mail: araujo@ieee.org

J.M. Araújo
Programa de Pós-Graduação em Engenharia Elétrica, Escola Politécnica da UFBA,
Universidade Federal da Bahia, Rua Aristides Novis, 02, Federação,
40210-630 Salvador, BA, Brazil

C.E.T. Dórea
Departamento de Engenharia de Computação e Automação,
Universidade Federal do Rio Grande do Norte, UFRN-CT-DCA,
59078-900 Natal, RN, Brazil
e-mail: cetdorea@dca.ufrn.br

© Springer International Publishing Switzerland 2016
H.E. Ponce Espinosa (ed.), *Nature-Inspired Computing for Control Systems*,
Studies in Systems, Decision and Control 40, DOI 10.1007/978-3-319-26230-7_6

1 Introduction

1.1 Overview

Currently, the design of control and estimation systems in the presence of constraints is a subject of great relevance. For instance, model predictive control (MPC) techniques often incorporate constraint-handling facilities [1–4]. Such constraints can represent control effort limitations, safe operational bounds on the process, maximum rate values, etc. Numerous studies describe techniques used in constrained systems, such as the anti-windup approach [5–8], set membership [9–12], and positively, controlled, and conditioned invariant sets [13–17]. The main advantage of the controlled and conditioned invariant sets approach is that it considers the constraints in the entire design. Nevertheless, some problems remain unresolved. In the static output feedback case, for a given invariant polyhedral set, an offline (analytical) control law has not been established for the general case [18]. In addition, for invariant estimators (observers) with multiple outputs, no direct method exists for designing offline output injection laws [19]. However, in both cases, the control inputs can be computed online by means of linear programming (LP) at each step [19, 18]. In this case, the disadvantage is the computational complexity inherent in the solution of a large number of LP problems. Higher-order systems or polyhedra with many faces or vertices entail a significant increase in the runtime for the control input computation. In this case, the optimal control input/output injections are piecewise affine (PWA) functions of the outputs/output errors. Recent work has demonstrated the high capability of multi-layer perceptron neural networks (MLP-NNs) in capturing these behaviors for adequate fitting of static output online feedback controllers [20] and online output injection laws in constrained error estimators [21]. MLP-NNs are known to be highly efficient in function approximation, offering constant-time computational complexity [22] and thus outperforming the most efficient lookup tables or search tree algorithms. The present work aims to apply the design of dynamic output feedback controllers to constrained linear systems. The objective is to use the combined design of constrained observers and feedback of the estimated state of the system, by using NNs to fit the input-output manifolds of those functions. To this end, simulated data captured from online observers and dynamic controllers are used as training sets for feedforward MLP-NNs. The runtime for the computation of the control effort is compared with that of the online strategy, and a drastic reduction in this parameter is demonstrated, thereby verifying the advantage of the approach.

1.2 Related Work

The concept of using NNs for the approximation of complex control structures in relevant applications has been explored in the specialized literature. These applications include MPC cost-function-based approximations [23], nonlinear MPC for *pH* process control [24], output feedback stabilization of interconnected nonlinear systems [25], and friction modeling and compensation [26]. In two recent studies, the authors and collaborators obtained preliminary results for the approximation of constrained controllers and observers designed via set invariance techniques [20, 21]. The present work bridges these two studies via approximation of dynamic output feedback controllers based on the set invariance approach.

2 Invariant Polyhedral Sets

2.1 Controlled Invariant Polyhedral Sets

Consider a linear, time-invariant, discrete-time system subject to unknown but bounded disturbance and measurement noise, expressed as

$$x(k+1) = Ax(k) + Bu(k) + Ed(k),$$
$$y(k) = Cx(k) + \eta(k), \tag{1}$$

where $x \in \mathscr{R}^n$ is the state, $u \in \mathscr{R}^p$ is the control input, $d \in \mathscr{R}^q$ is the disturbance, $y \in \mathscr{R}^l$ is the measured output, $\eta \in \mathscr{R}^l$ is the measurement noise, and k is the sampling time. Matrices B, E, and C are full rank matrices.

The disturbance d and the measurement noise η are assumed to be unknown but bounded to compact convex sets (C-sets) $\mathscr{D} \subset \mathscr{R}^q$ and $\mathscr{N} \subset \mathscr{R}^l$, respectively. Moreover, the system is subject to control constraints $u \in \mathscr{U}$, where $\mathscr{U}$ is also a C-set.

Consider a closed convex set $\Omega \subset \mathscr{R}^n$ whose interior contains the origin.

Definition 1 The set $\Omega \subset \mathscr{R}^n$ is said to be *controlled invariant* with respect to (w.r. t.) system (1) if $\forall x \in \Omega$, $\exists u \in \mathscr{U}$ such that $Ax + Bu + Ed \in \Omega$, and the measurement noise $\forall d \in \mathscr{D}$.

Clearly, if Ω is controlled invariant, there exists a state feedback $u(x(k))$ such that $x(k) \in \Omega$ $\forall k$, for all admissible disturbances $d(k) \in \mathscr{D}$.

Very often, a *contraction rate* $0 < \lambda < 1$ is imposed on the controlled invariant set so as to guarantee convergence of the state to the origin (or to a small set around the origin if persistent disturbances are present). In this case, the invariance condition in the above definition is given by $Ax + Bu + Ed \in \lambda\Omega$.

In this work, state and control constraints as well as disturbances and measurement noise are assumed to belong to convex polyhedral sets as follows:

$$\Omega = \{x : Gx \le \rho\}$$
$$\mathscr{U} = \{u : Uu \le v\}$$
$$\mathscr{D} = \{d : Dd \le \omega\} \tag{2}$$
$$\mathscr{N} = \{\eta : |\eta| \le \bar{\eta}\}$$

In general, the polyhedral set Ω, which defines the constraints on the state trajectory, is not controlled invariant. Nevertheless, it is possible to confine the state trajectory to a controlled invariant set contained in Ω. The maximal controlled invariant set contained in Ω can be characterized, and numerical algorithms for its computation are available [13, 15]. More recent efficient methods can be used if the maximal set is not required [27].

Once an admissible controlled invariant polyhedral set is available, the sequence of control inputs that enforces the constraints can be computed online via linear programming (LP) or offline by computing a piecewise linear state feedback control law [13] or a piecewise affine control law via multi-parametric linear programming [28]. However, the complexity of such laws increases rapidly with the system dimension. Therefore, it is very difficult to implement them in practice. Alternatives to such laws have been sought, one of which involves the use of NNs to emulate them.

2.2 Conditioned Invariant Polyhedral Sets

Consider the set of *admissible outputs* associated with Ω:

$$\mathscr{Y}(\Omega) = \{y : y = Cx + \eta \text{ for } x \in \Omega, \eta \in \mathscr{N}\}. \tag{3}$$

$\mathscr{Y}(\Omega) \subset \mathscr{R}^l$ is the set of all values of y that can be associated with $x \in \Omega$. Therefore, if $x(k) \in \Omega$, then $y(k) \in \mathscr{Y}(\Omega)$.

With each $y \in \mathscr{R}^l$ is associated the following set of states consistent with the measurement:

$$\mathscr{C}(y) = \{x : Cx = y - \eta, \text{ for } \eta \in \mathscr{N}\}. \tag{4}$$

Definition 2 *The set $\Omega \subset \mathscr{R}^n$ is said to be conditioned invariant w.r.t. system (1) if* $\forall y \in \mathscr{Y}(\Omega), \exists v : Ax + v + Ed \in \Omega, \forall d \in \mathscr{D}$ *and* $\forall x \in \Omega, \eta \in \mathscr{N} : Cx + \eta = y$.

Considering a contraction rate, the conditioned invariance condition is given by $Ax + v + Ed \in \lambda\Omega$.

Definition 2 can be rewritten in geometric terms as follows: Ω is conditioned invariant if $\forall y \in \mathscr{Y}(\Omega), \exists v : A[\mathscr{C}(y) \cap \Omega] + v + E\mathscr{D} \subset \Omega$.

Conditioned invariant sets have been used in [19, 29, 30, 31] to construct *set-invariant estimators*, i.e., state observers for which the estimation error is bounded by a conditioned invariant set. Numerical methods for computing the minimal (or as small as possible) conditioned invariant polyhedron that contains a given polyhedron can also be found therein. Typically, the given polyhedron is defined by the uncertainty in the initial estimation error. The conditioned invariant polyhedron is intended to bound as much as possible the trajectory of the estimation error. An extension of this concept to descriptor linear systems has been described in [29], along with its application to a three-tank system.

In this framework of observer design, the state space model is written in terms of the estimation error as follows:

$$
\begin{aligned}
e(k+1) &= Ae(k) + v(k) + Ed(k), \\
z(k) &= Ce(k) + \eta(k),
\end{aligned}
\tag{5}
$$

where $e(k)$ is the state estimation error and $z(k)$ is the output estimation error. The vector $v(k)$ denotes the output injection.

Once an admissible conditioned invariant polyhedral set is available, the sequence of output injections $v(k)$ that enforces the invariance of Ω can also be computed online via LP. Let $\Omega = \{x : Gx \leq \rho\}$ be conditioned invariant w.r.t. (1). Then, according to Definition 2, if $y \in \mathscr{Y}(\Omega)$, there exists v such that $Ax + v + Ed \in \Omega$, $\forall d \in \mathscr{D}$ and $\forall x \in \Omega$, such that $Cx + \eta = y$ for some $\eta \in \mathscr{N}$. Considering the polyhedral sets defined in (2), the condition $Ax + v + Ed \in \Omega$ becomes

$$
G(Ax + v + Ed) \leq \rho, \forall x \text{ such that } Gx \leq \rho, |Cx - y| \leq \bar{\eta}.
$$

A single v should work for all the inequalities above; thus, the worst-case x can be considered row by row. Consider the vector $\phi(y)$ whose components are given by

$$
\phi_i(y) = \max_x G_iAx, \text{ subject to: } Gx \leq \rho, |Cx - y| \leq \bar{\eta}.
\tag{6}
$$

In addition, consider the vector δ, which accounts for the influence of disturbances, whose components are given by

$$
\delta_i = \max_d GEd \text{ subject to: } Dd \leq \omega.
$$

Then, the vector v can be computed online by the solution of the following LP:

$$
\begin{aligned}
&\min_{\varepsilon, v(k)} \varepsilon \\
&\text{s.t. } \phi(y) + Gv \leq \varepsilon\rho - \delta.
\end{aligned}
\tag{7}
$$

The online computation of $v(k)$ is carried out as follows. For each time step k, solve g LPs (6) in order to compute $\phi_i(y(k))$, where g is the number of rows in G. Then, solve the LP (7). Keeping in mind that the number of rows in G increases rapidly with the system dimension, it is easy to observe that such an online computation can hardly be implemented in practical problems.

An offline piecewise affine output injection law has been proposed in [30] for single-output systems. In [19], it was shown that a piecewise affine can be computed for multiple-output systems via multi-parametric LP, but with very high complexity. In [21], an NN that will be discussed later was designed to emulate the output injection.

2.3   Output Feedback Controlled Invariant Polyhedra

The use of controlled invariant polyhedra for constrained control assumes full measurement of the state, which is seldom possible in practice owing to technical or cost limitations. Thus, the output feedback case is preferred.

Consider the set $\mathscr{Y}(\Omega)$ of all admissible outputs associated with Ω (3) and the set $\mathscr{C}(y)$ of states consistent with a given measurement y (4). The following concept characterizes controlled invariance via output feedback [18]:

Definition 3 *The polyhedron* Ω *is* output feedback controlled invariant (o.f.c.i.) *w.r.t. system* (1) *if* $\forall y \in \mathscr{Y}(\Omega)$, $\exists u(y) \in \mathscr{U} : G(Ax + Bu + Ed) \le \lambda\rho$, $\forall d \in \mathscr{D}$ *and* $\forall \mathbf{x} \in \Omega\cap\mathscr{C}(y)$, *with* $0 < \lambda \le 1$.

In other words, if the state at time k belongs to Ω, with Ω o.f.c.i, then, from the knowledge of $y(k)$, it is possible to enforce $x(k+1) \in \lambda\Omega$ through the computation of $u(k) \in \mathscr{U}$, in spite of the disturbances $d(k) \in \mathscr{D}$ and noise $\eta(k) \in \mathscr{N}$. Therefore, if the initial state $x(0)$ belongs to Ω, then by means of a suitable static output feedback $u(y(k))$, it is possible to keep $x(k)$ in Ω $\forall k$.

Definition 3 can be rewritten in geometric terms as follows: Ω is o.f.c.i. if $\forall y \in \mathscr{Y}(\Omega)$, $\exists u \in \mathscr{U} : A[\mathscr{C}(y)\cap\Omega] + Bu + E\mathscr{D} \subset \Omega$.

A straightforward consequence of Definitions 1–3 is that Ω is o.f.c.i only if it is simultaneously controlled and conditioned invariant.

The conditions under which a given polyhedron is o.f.c.i. have been established in [18]. However, as opposed to the case of controlled and conditioned invariant polyhedra, there is no method for directly computing o.f.c.i. polyhedral sets. If we restrict ourselves to static output feedback, there are two possibilities. First, if the open-loop system is asymptotically stable, then an invariant polyhedron can be computed w.r.t. the autonomous system. Clearly, such a polyhedron is o.f.c.i. as well. Second, if the open-loop system is unstable but stabilizable via output feedback, a stabilizing control can be designed, and an invariant set can be computed w.r.t. the closed-loop system. However, both possibilities indicate some conservatism.

Another possibility is the use of dynamic output feedback. Consider system (1) and the following (possibly nonlinear) compensator:

$$z(k+1) = v(z(k), y(k)), \tag{8}$$

$$u(k) = \kappa(z(k), y(k)). \tag{9}$$

System (1) under the compensator (9) can be represented in an extended state space formulation as follows:

$$\xi(k+1) = \hat{A}\xi(k) + \hat{B}\omega(k) + \hat{E}d(k), \tag{10}$$

$$\zeta(k) = \hat{C}\xi(k) + \hat{\eta}(k), \tag{11}$$

where $\xi = \begin{bmatrix} x \\ z \end{bmatrix}$ is the extended state, $\omega = \begin{bmatrix} u \\ v \end{bmatrix}$ is the extended input vector, $\zeta = \begin{bmatrix} y \\ z \end{bmatrix}$ is the extended output vector, $\hat{\eta} = \begin{bmatrix} \eta \\ 0 \end{bmatrix}$ is the extended noise vector, $\hat{A} = \begin{bmatrix} A & 0 \\ 0 & 0 \end{bmatrix}$, $\hat{B} = \begin{bmatrix} B & 0 \\ 0 & I \end{bmatrix}$, $\hat{E} = \begin{bmatrix} E \\ 0 \end{bmatrix}$, and $\hat{C} = \begin{bmatrix} C & 0 \\ 0 & I \end{bmatrix}$. Further, $u(k)$ and $v(k)$ are functions of the extended output vector $\begin{bmatrix} y(k) \\ z(k) \end{bmatrix}$, as expressed in (9).

Control constraints and bounds on the measurement noise are given by
$$\omega(k) \in \mathscr{U} = \left\{ \omega = \begin{bmatrix} u \\ v \end{bmatrix} : [U \quad 0]\begin{bmatrix} u \\ v \end{bmatrix} \le v \right\}, \hat{\eta} \in \mathscr{N} = \left\{ \hat{\eta} = \begin{bmatrix} \eta \\ 0 \end{bmatrix} : |\eta| \le \bar{\eta} \right\}.$$

Now, consider a pair of compact convex polyhedral sets $(\mathscr{S}, \mathscr{V})$, given by

$$\mathscr{S} = \{x : G_s x \le 1\}, \mathscr{V} = \{x : G_v x \le 1\},$$

and satisfying the following assumptions:

Assumption $\mathscr{S} \subset \mathscr{V} \subset \Omega$; $\mathscr{S}$ is conditioned invariant and $\mathscr{V}$ is controlled invariant.

The following result, first presented in [18], can be established.

Proposition 1 *The polyhedron*

$$\hat{\mathscr{P}} = \{\xi : \hat{G}\xi \le 1\}, \text{ with } \hat{G} = \begin{bmatrix} G_v & 0 \\ G_s & -G_s \end{bmatrix}, \tag{12}$$

is simultaneously controlled and conditioned invariant w.r.t. system (11).

Then, the polyhedron $\hat{\mathscr{P}}$ is a candidate to be o.f.c.i. w.r.t. the extended system (11), because it satisfies the necessary conditions. The conditions proposed in [18] can be used to check if it is o.f.c.i.

A pair $(\mathscr{S}, \mathscr{V})$ satisfying the assumption stated above can be obtained from the available algorithms for the computation of controlled and conditioned invariant polyhedra mentioned above. A natural choice for $\mathscr{V}$ would be the maximal controlled invariant set contained in Ω. For the conditioned invariant set $\mathscr{S}$, if the initial state belongs to a *confidence set*, then $\hat{\mathscr{S}}$ can be chosen to be the minimal (or as small as possible) conditioned invariant set containing it.

Extensions of output feedback controlled invariance computation to linear descriptor systems have been proposed in [32] (static output feedback) and [33] (dynamic output feedback).

The control input that guarantees constraint satisfaction as stated in Definition 3 can be computed online at each step as follows [18]:

$$u(y(k)) = \arg \min_{\varepsilon, u} \varepsilon$$
$$\text{s.t. } \phi(y(k)) + GBu \leq \varepsilon\rho - \delta, \quad Uu(k) \leq v. \tag{13}$$

This control input tries to minimize the one-step contraction rate of the state trajectory. The terms ϕ and δ denote the worst-case admissible states consistent with a given measure and the worst-case disturbance action. Both involve LP in their computation. The term δ is computed only once, but ϕ must be computed online at each step. Thus, the computational effort increases with the size of the invariant polyhedron or the system order; a long runtime can make it difficult to control systems with fast dynamics.

3 Neural Network Multilayer Perceptrons

3.1 Natural Inspiration of the Work

Artificial Neural Networks—or simple Neural Networks (NNs) are engineered computational systems that mimetize at certain level the learning capacity and the parallelism of human brain, reducing thereby the gap between that and Turing Machines [34, 35]. Among the several tecnhiques of the class of computational intelligence, NNs were one of the first that have been introduced, being clearly a nature inspired or biomimicry approach [36]. The following subsections cover some relevant information and references about Neural Network Multilayer Perceptrons (NN-MLPs).

3.2 Neural Networks for Fitting Function

Multi-layer perceptron neural networks (MLP-NNs or just NNs) constitute a family of techniques in the broader research area known as computational intelligence. NNs are mathematical structures that include a mechanism for learning from input and output data. Since the initial development of NNs several decades ago [37, 38], several NN applications have been consolidated: control and modeling of dynamical systems, function approximation, pattern recognition, and time-series prediction [39–48]. In addition, the learning techniques that constitute the field known as machine learning are key features of NN structures, and several contributions have facilitated the development of general and specialized frameworks [49–54].

Fig. 1 Signal flow graph for one perceptron

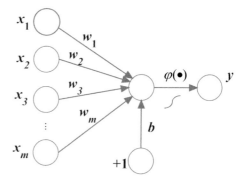

Figure 1 shows a flow graph for a single neuron or perceptron, and Fig. 2 shows the classical architecture of MLPs. The output layer functions is linear in general (Σ), however, the designer may adopt other functions. The transfer function of a single perceptron is given by

$$y = \varphi\left(\sum_{i=1}^{m} w_i x_i + b\right),\tag{14}$$

where w_i is the ith input weight, b is the constant bias, and $\varphi(\bullet)$ is the so-called activation function—generally, a sigmoid hyperbolic tangent or logistic function.

In particular, function approximation using NNs is one of the most well-developed applications. The first relevant contribution to this theme was presented in [42]. Among other issues, it has been shown in this work that any multivariate function can be arbitrarily approximated in a unit hypercube support domain with a single hidden layer feedforward NN consisting of sigmoid activation

Fig. 2 Signal flow graph for an MLP

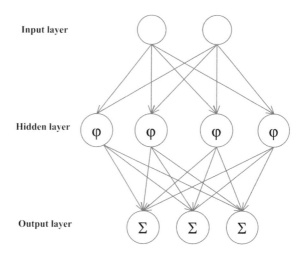

functions. Other relevant contributions include the equivalence of MLP-NNs with
networks based on Gaussian potential functions (GPFs) [55], function approximation training based on an algebraic approach [56], improvements in the
approximation of non-smooth (piecewise affine) functions [57], and more recent
advances in function approximations with NNs based on sigmoid functions [58].

The focus of this work is the function approximation capability of NNs.
Modeling functions with feedforward NNs are highly attractive because the computational complexity of trained networks for evaluating the approximant is of
constant-time type, as opposed to the complexity of lookup tables and search trees,
i.e., $\Theta(\log n)$ [59]. A series of contributions to the implementation and application
of function approximation and other related applications of NNs can be found in the
specialized literature [60–65]. The manifolds of functions (7) and (13) are known to
exhibit PWA behavior. The next section provides an outline of the proposed
methodology, which can be applied to NNs to approximate static and dynamic
output feedback controllers for constrained linear systems.

4 Proposed Approach

The methodology for adequate application of NNs to the design of static and
dynamic controllers entails generating cloud points of input-output pairs by simulating the online schemes (7) or (13). A grid of points is obtained from the
invariant polyhedra in order to give a consistent set of initial conditions that feed the
simulations, generating the training sets of the NNs. Unlike [18], here, in
the dynamic output feedback case, the estimated state from the observer was used as
the output, i.e., $y_1(k) = f(y(k)) = \hat{x}(k)$. Thus, the new output equation for (1) is
given by

$$y_1(k) = x(k) - e(k). \tag{15}$$

Hence, a given invariant polyhedron obtained, e.g., using the algorithms presented in [15] can be checked for o.f.c.i. by taking $C = I$ and $\eta(k) = -e(k)$.

The details of the feedforward NN characteristics—layers, activation functions,
training algorithm, etc.—are provided in the next section.

5 Results and Discussion

In order to illustrate the features of the proposed approach, two examples are
presented. The first example deals with a constrained static output feedback design
in a controllable triple-integrator type plant. The second example deals with the
simulation of a real-world system that involves a two-tank liquid level constrained

regulation problem. For all cases, the data set was divided into training (70 %), test (15 %), and validation (15 %) sets.

5.1 Triple Integrator Constrained Control

Consider a controllable, undisturbed triple integrator plant with output measurement noise:

$$A = \begin{bmatrix} 0 & 1 & 0 \\ 0 & 0 & 1 \\ 0 & 0 & 0 \end{bmatrix}, B = \begin{bmatrix} 0 \\ 0 \\ 1 \end{bmatrix}, C = \begin{bmatrix} 1 & 0 & 0 \\ 0 & 2 & 2 \end{bmatrix}$$

The following constraints are given for the system:

$$|x_i| \leq 1, i = 1, 2, 3; |u| \leq 10; |\eta| \leq 0.01$$

This system was discretized with sample time $T_s = 0.1$ s, and two designs, one for static output feedback and the other for dynamic output feedback, were obtained as follows.

The static output feedback design was obtained in the following steps:

1. A controlled invariant polyhedron with contraction rate $\lambda = 0.96$ was computed using the algorithm proposed in [15].
2. This polyhedron was checked to be o.f.c.i. with maximum contraction rate $\lambda_{max} = 1$.

A random set of initial conditions was generated, including the vertices of the polyhedron. Then, a training set for an MLP-NN was obtained by saving the output trajectories and the corresponding control inputs computed according to (13). The neural controller was implemented using an MLP-NN consisting of one hidden layer having 33 neurons, with linear saturated (satlin) activation functions at the hidden and output layer; the training algorithm used was the Levenberg-Marquardt algorithm. The trained NN exhibited good performance in fitting the input-output manifold of the control law, with a mean square error (MSE) of $2.3e-4$. Figure 3 shows the input-output manifold of the neural controller.

For the dynamic output feedback case, the procedure described in Sect. 3 was adopted. First, an observer was designed with the following constraint specifications:

$$|e_i| \leq 0.05, \ i = 1, 2, 3; \quad |\eta| \leq 0.01.$$

If the system has multiple outputs, it is difficult to construct an analytical observer. A random set of initial conditions was used for simulation of the online observer in order to collect the necessary data for training, testing, and validating

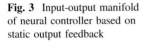

Fig. 3 Input-output manifold
of neural controller based on
static output feedback

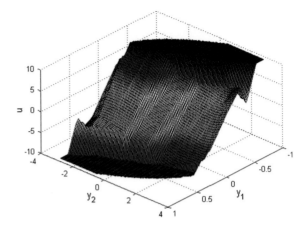

the neural observer. The invariant polyhedron for this case is shown in Fig. 4. The
neural observer was implemented using an MLP-NN having 19 neurons at the
unique hidden layer with symmetric saturating linear (satlins) activation functions;
the training algorithm used was the Levenberg-Marquardt algorithm. The trained
network exhibited excellent performance for fitting, with an *MSE* of $7.75e - 8$. The
input-output manifold for the observer is shown in Fig. 5.

Considering the feedback with the estimated state from such an observer, the
polyhedron was checked to be o.f.c.i. with a contraction rate $\lambda_{max} = 0.9716$, which
was better than that of the static feedback case. Again, an MLP-NN was trained for
fitting the map between the estimated state $\hat{x}$ and the control effort u. This neural
controller was implemented using an MLP-NN having 35 neurons at the hidden
layer with sigmoidal hyperbolic tangent (tansig) activation functions; the training
algorithm used was the Levenberg-Marquardt algorithm. The trained network
achieved satisfactory performance, with an *MSE* of 0.11.

Fig. 4 Invariant polyhedron
for the triple integrator
observer

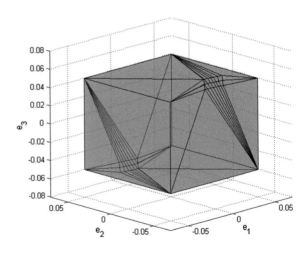

Fig. 5 Manifold for the
multi-input observer of the
triple integrator

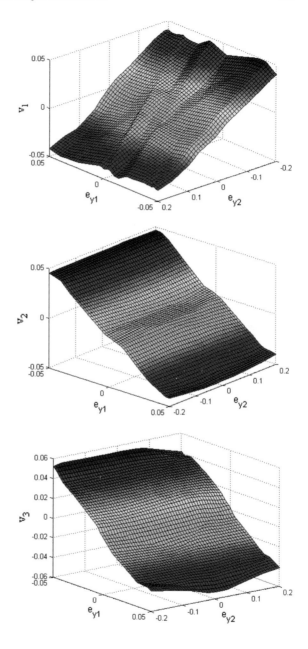

Figures 6 and 7 compare the performances of static and dynamic feedback-based neural controllers. The invariant polyhedron for this case is shown in Fig. 8, along with a trajectory starting at the border of the polyhedron.

Fig. 6 Time-domain state
trajectory for neural
controllers: static (*dash line*)
and dynamic (*solid line*), and
the colors are for the states x_1
(*blue*), x_2 (*red*) and x_3 (*green*)

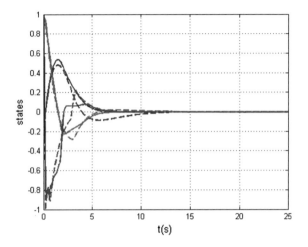

Fig. 7 Control effort for
neural controllers: static
(*dashed line*) and dynamic
(*solid line*)

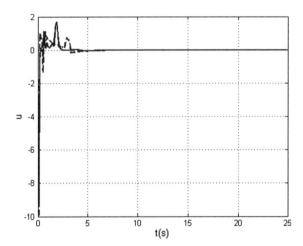

Fig. 8 Invariant polyhedron
with state trajectories for
static and dynamic
feedback-based neural
controllers: static (*red line*)
and dynamic (*blue line*)

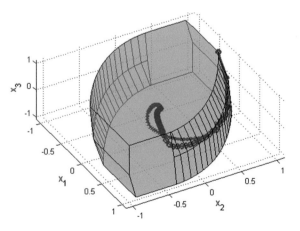

Fig. 9 Cascade hydraulic
tank system

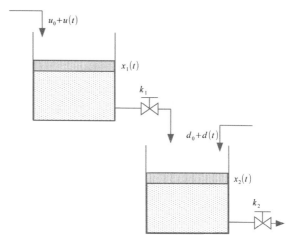

5.2 Constrained Regulation in a Hydraulic System

The system shown in Fig. 9 consists of two uncoupled cascade tanks, with the
control being the feeding flow into the upper tank. The state-space system model is
given by

$$\dot{x}(t) = \begin{bmatrix} -k_1 & 0 \\ k_1 & -k_2 \end{bmatrix} x(t) + \begin{bmatrix} 1 \\ 0 \end{bmatrix} u(t) + \begin{bmatrix} 0 \\ 1 \end{bmatrix} d(t)$$

$$y(t) = [\,1 \quad -1\,] x(t) + \eta(t).$$

The parameters used in the simulation were $k_1 = 3\,\mathrm{s}^{-1}$, $k_2 = 1\,\mathrm{s}^{-1}$, and
$T_s = 0.05\,\mathrm{s}$. The constraints considered for invariant polyhedra computations were
$|x_1| \leq 5, |x_2| \leq 5, |u| \leq 5, |d| \leq 0.1$, and $|\eta| \leq 0.05$,

Using the algorithm proposed in [15], an invariant polyhedron with contraction
rate $\lambda = 0.9$ for the full state feedback case was obtained. This set is shown in
Fig. 10. The number of facets is 48, indicating a reasonable computational effort for
the online control computation.

Three situations were explored in the application of NNs.

5.3 Static Output Feedback

Using the methodology described in [18], the computed polyhedron was confirmed
to be o.f.c.i. with the maximum contraction rate $\lambda_{max} = 0.967$. A training set for the

Fig. 10 Invariant polyhedron
for the cascade hydraulic
system

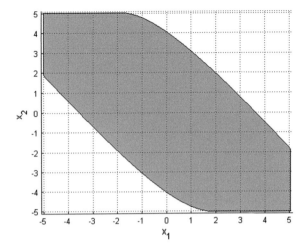

Fig. 11 Grid of initial
conditions for system
simulation

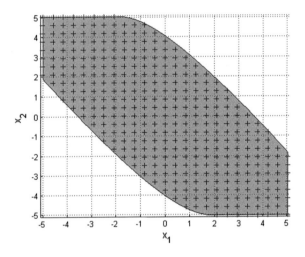

NN was obtained by simulating the trajectories starting from the initial conditions in the grid shown in Fig. 11. After some tests, an adequate NN was trained with the following characteristics: one hidden layer with 39 neurons; the activation function was symmetric saturated linear; the output function was linear; and the training algorithm used was the Levenberg-Marquardt algorithm. After 32 epochs of training, the NN exhibited an *MSE* of $4.87e-6$. Figure 12 shows the input-output manifold of the NN. One can easily identify a PWA profile, despite the apparently smooth curvature in the central region, which is due to the large number of facets of the invariant polyhedron.

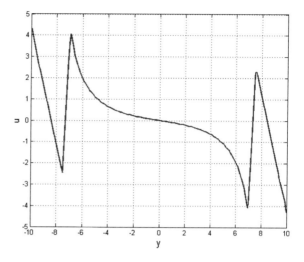

Fig. 12 Input-output manifold map for neural static controller in the cascade tank configuration

5.4 Observer with Error Limitation

For this design, a contraction rate of $\lambda = 0.8$ was established for the observer. An initial confidence set for the error was taken as $|e_1| \leq 0.5$ and $|e_2| \leq 0.5$. The disturbance and measurement noise are the same as those in the static output feedback case. Again, the initial conditions used to generate the training set of the NN observer were those in the grid of Fig. 11. The minimal conditioned invariant polyhedron, which contains the original constraint set, was computed using the algorithms presented in [29, 66]; it is shown in Fig. 13. The NN for this case was structured with one hidden layer containing 11 neurons; the activation function was symmetric saturated linear; the output layer was linear; and the training algorithm

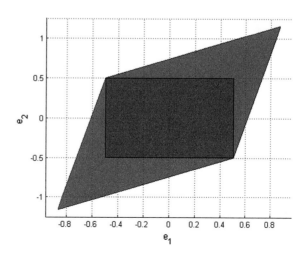

Fig. 13 Conditioned invariant polyhedron and original constraints in the observer design

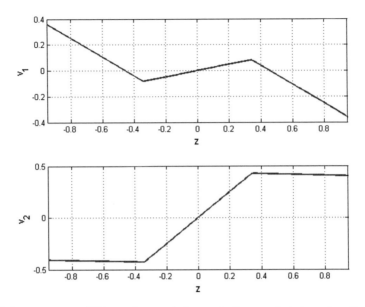

Fig. 14 Input-output manifold map for neural observer in the cascade tank configuration

used was the Levenberg-Marquardt algorithm. After 12 epochs of training, the NN exhibited satisfactory performance, with an *MSE* of 4.06e−8. The input-output manifold is shown in Fig. 14, again with a clear PWA profile.

5.5 Dynamic Output Feedback

In this case, the designed observer was employed for feedback, and the invariant polyhedron recovered its original contraction rate using the output (15) to provide feedback. The NN used here was structured with 27 neurons at the hidden layer; activation of the hyperbolic tangent function at the hidden layer was the best choice, giving a smaller *MSE*; a symmetric saturated linear function was obtained at the output layer; and the training algorithm used was the Levenberg-Marquardt algorithm. Owing to the inherent difficulty in approximating functions of more than one variable, the NN for this case was retrained a few times, and suitable training was finally achieved after 57 epochs, with an *MSE* of 3.3e−3. Figure 15 shows the color-scaled input-output manifold projected onto the estimated state space of the controller.

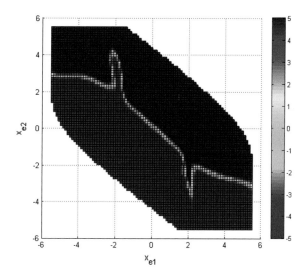

Fig. 15 Input-output manifold map for dynamic neural controller in the cascade tank configuration

5.6 Discussion on Runtime and Performance Comparison

The runtime of the NNs was estimated by simulating the obtained controller in MATLAB© on a system with $Intel(R)Core(TM)i7 - 3517UCPU@1.90$ GHz. Table 1 compares the runtime of online computation with that of the NNs. In all cases, the runtimes of the NNs were considerably lower than those of online computation. The NN performance is essentially invariant in both examples, despite the different orders of the systems. In addition, the runtime of the dynamic neural controller was roughly twice as long as that of the static neural controller; this can be explained by the use of two NNs to perform the control input computation.

Figure 16 shows that the NN exhibits better performance in the dynamic output feedback case than in the static output feedback case; the dynamic controller (solid line) achieves faster regulation of liquid levels than the static one (dashed line). The control effort for both cases is shown in Fig. 17, which confirms that the dynamic output feedback case has a better state response.

Table 1 Runtime summary for the studied examples

Example		Online controller (μs)	Neural controller (μs)
Triple integrator	Static output feedback	12,000	160
	Dynamic output feedback	20,000	330
Hydraulic system	Static output feedback	5200	165
	Dynamic output feedback	5300	320

Fig. 16 Time-domain
performance comparison of
neural controllers: static
(*dashed line*) and dynamic
(*solid line*)

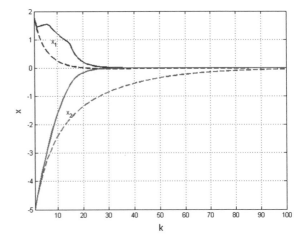

Fig. 17 Control effort of
neural controllers: static
(*dashed line*) and dynamic
(*solid line*)

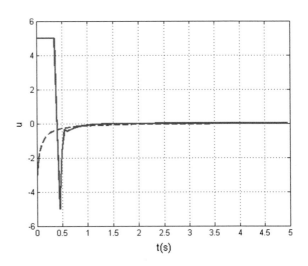

6   Concluding Remarks and Future Work

This work proposed a new approach for control and estimation in constrained linear
systems using MLP-NNs. Two numerical examples were presented, in which the
neural controllers and observers intended to replace online computation exhibited
comparable performance and better runtimes. These results, together with the
present state of NN implementation using fast hardware, indicate the high potential
of the proposed approach for real-time and critical applications of constrained
control and estimation design. Future efforts must be devoted to experimental
evaluation of the proposed approach using FPGA platforms.

References

1. Kothare, M.V., Balakrishnan, V., Morari, M.: Robust constrained model predictive control using linear matrix inequalities. Automatica **32**(10), 1361–1379 (1996)
2. Kothare, S.L.D.O., Morari, M.: Contractive model predictive control for constrained nonlinear systems. IEEE Trans. Autom. Control **45**(6), 1053–1071 (2000)
3. Mayne, D.Q., Rawlings, J.B., Rao, C.V., Scokaert, P.O.M.: Constrained model predictive control: stability and optimality. Automatica **36**(6), 789–814 (2000)
4. Mayne, D.Q., Seron, M.M., Raković, S.V.: Robust model predictive control of constrained linear systems with bounded disturbances. Automatica **41**(2), 219–224 (2005)
5. Hanus, R., Kinnaert, M., Henrotte, J.L.: Conditioning technique, a general anti-windup and bumpless transfer method. Automatica **23**(6), 729–739 (1987)
6. Kothare, M.V., Campo, P.J., Morari, M., Nett, C.N.: A unified framework for the study of anti-windup designs. Automatica **30**(12), 1869–1883 (1994)
7. Zaccarian, L., Teel, A.: Modern Anti-windup Synthesis: Control Augmentation for Actuator Saturation: Control Augmentation for Actuator Saturation. Princeton Series in Applied Mathematics. Princeton University Press (2011)
8. Zheng, A.: Mayuresh, Kothare, V., Morari, M.: Anti-windup design for internal model control. Int. J. Control **60**(5), 1015–1024 (1994)
9. Haimovich, H., Goodwin, G., Welsh, J.: Set-valued observers for constrained state estimation of discrete-time systems with quantized measurements. In: 2004. 5th Asian Control Conference, vol. 3, pp. 1937–1945 (2004)
10. Keesman, K.J., Stappers, R.: Nonlinear set-membership estimation: a support vector machine approach. J. Inverse Ill-posed Probl. **12**(1), 27–42 (2004)
11. Raïssi, T., Ramdani, N., Candau, Y.: Set membership state and parameter estimation for systems described by nonlinear differential equations. Automatica **40**(10), 1771–1777 (2004)
12. Shamma, J., Tu, K.Y.: Set-valued observers and optimal disturbance rejection. IEEE Trans. Autom. Control **44**(2), 253–264 (1999)
13. Blanchini, F.: Ultimate boundedness control for uncertain discrete-time systems via set-induced lyapunov functions. IEEE Trans. Autom. Control **39**(2), 428–433 (1994)
14. Darup, M.S., Mönnigmann, M.: A stabilizing control scheme for linear systems on controlled invariant sets. Syst. Control Lett. **79**, 8–14 (2015)
15. Dórea, C.E.T., Hennet, J.C.: (A, B)-invariant polyhedral sets of linear discrete-time systems. J. Optim. Theory Appl. **103**(3), 521–542 (1999)
16. Raković, S., Kerrigan, E.C., Mayne, D.Q., Kouramas, K.I.: Optimized robust control invariance for linear discrete-time systems: Theoretical foundations. Automatica **43**(5), 831–841 (2007)
17. Vassilaki, M., Bitsoris, G.: Constrained regulation of linear continuous-time dynamical systems. Syst. Control Lett. **13**, 247–252 (1989)
18. Dórea, C.E.T.: Output-feedback controlled-invariant polyhedra for constrained linear systems. In: Proceedings of 48th Concefernce in Decision and Control, pp. 5317–5322 (2009)
19. Dorea, C.E.T.: Set-invariant estimators for multiple-output discrete-time systems. In: 2006 45th IEEE Conference on Decision and Control, pp. 4538–4543 (2006)
20. do Carmo, S., de Almeida, M., de Castro, F., Campos, R., Araujo, J., Dórea, C.: Neural network fitting for input-output manifolds of online control laws in constrained linear systems. In: 2014 IEEE Symposium on Computational Intelligence in Control and Automation (CICA), pp. 1–6 (2014)
21. Almeida, M., Carmo, S., Castro, F., Campos, R., Araújo, J., Dórea, C.: Neural network fitting for input-output manifolds of online observers with error limitation. In: 2014 Brazilian Congress on Automatica (2014)
22. Parri, J., Ratti, S.: Trigonometric function approximation neural network based coprocessor. In: Microsystems and Nanoelectronics Research Conference, 2009. MNRC 2009. 2nd, pp. 148–151 (2009)

23. Akesson, B.M., Toivonen, H.T.: A neural network model predictive controller. J. Process Control **16**(9), 937–946 (2006)
24. Akesson, B.M., Toivonen, H.T., Waller, J.B., Nyström, R.H.: Neural network approximation of a nonlinear model predictive controller applied to a pH neutralization process. Comput. Chem. Eng. **29**(2), 323–335 (2005)
25. Hua, C., Guan, X.: Output feedback stabilization for time-delay nonlinear interconnected systems using neural networks. IEEE Trans. Neural Netw. **19**(4), 673–688 (2008)
26. Huang, S., Tan, K.K.: Intelligent friction modeling and compensation using neural network approximations. IEEE Trans. Ind. Electron. **59**(8), 3342–3349 (2012)
27. Raković, S., Baric, M.: Parameterized robust control invariant sets for linear systems: Theoretical advances and computational remarks. IEEE Trans. Automat. Contr. **55**(7), 1599–1614 (2010)
28. Borrelli, F., Bemporad, A., Morari, M.: Geometric algorithm for multiparametric linear programming. J. Optimiz. Theory Appl. **103**(3), 521–542 (2003)
29. Araujo, J.M., Barros, P.R., Dóorea, C.E.T.: Design of observers with error limitation in discrete-time descriptor systems: a case study of a hydraulic tank system. IEEE Trans. Control Syst. Technol. (99), 1–7 (2011)
30. Dórea, C.E.T., Pimenta, A.: Design of set-invariant estimators for linear discrete-time systems. In: Proceedings of the 44th IEEE Conference on Decision and Control, pp. 7235–7240 (2005)
31. Dórea, C.E.T., Pimenta, A.: Set-invariant estimators for linear systems subject to disturbances and measurement noise. In: Proceedings of 16th IFAC World Congress (2005)
32. Araújo, J.M., Barroso, H.C., Barros, P.R., Dórea, C.E.T.: Output feedback control of constrained descriptor systems: a case study of a hydraulic tank system. Proc. Inst. Mech. Eng. Part I: J. Syst. Control Eng. **226**(3), 429–436 (2012)
33. Araújo, J.M., Barros, P.R., Dórea, C.E.T.: Dynamic output feedback control of constrained descriptor systems. Trans. Inst. Measure. Control **35**(8), 1129–1138 (2013)
34. Haykin, S.: Neural Networks: A Comprehensive Foundation. International edition, Prentice Hall International (1999)
35. Miller, W., Werbos, P., Sutton, R.: Neural Networks for Control. A Bradford book. MIT Press (1995). URL https://books.google.com.br/books?id=prjMtIr_yT8C
36. Passino, K.: Biomimicry for Optimization, Control, and Automation. Springer, London (2005)
37. Farley, B., Clark, W.: Simulation of self-organizing systems by digital computer. Trans. IRE Prof. Group Inf. Theory **4**(4), 76–84 (1954)
38. McCulloch, W., Pitts, W.: A logical calculus of the ideas immanent in nervous activity. Bull. Math. Biophys. **5**(4), 115–133 (1943)
39. Bishop, C.: Neural Networks for Pattern Recognition. Neural Networks for Pattern Recognition. Oxford University Press, USA (1995)
40. Chen, C.L.P., LeClair, S.R., Pao, Y.: An incremental adaptive implementation of functional-link processing for function approximation, time-series prediction, and system identification. Neurocomputing **18**(1–3), 11–31 (1998)
41. Chuang, C., Su, S., Chen, S.: Robust TSK fuzzy modeling for function approximation with outliers. IEEE Trans. Fuzzy Syst. **9**(6), 810–821 (2001)
42. Cybenko, G.: Approximation by superpositions of a sigmoidal function. Mathematics of Control, Signals, and Systems **2**(4), 303–314 (1989). Cited By 3688
43. Gorinevsky, D.: Adaptive learning control using affine radial basis function network approximation. In: Proceedings of the 1993 IEEE International Symposium on Intelligent Control, pp. 505–510 (1993)
44. Jones, R., Lee, Y., Barnes, C., Flake, G., Lee, K., Lewis, P., Qian, S.: Function approximation and time series prediction with neural networks. In: 1990 IJCNN International Joint Conference on Neural Networks, vol. 1, pp. 649–665 (1990)
45. Linse, D.J., Stengel, R.F.: Neural networks for function approximation in nonlinear control. In: Proceedings of the American Control Conference, pp. 674–679 (1990)

46. Martinetz, T.M., Berkovich, S.G., Schulten, K.J.: 'neural-gas' network for vector quantization and its application to time-series prediction. IEEE Trans. Neural Netw. **4**(4), 558–569 (1993). Cited By 778
47. Narendra, K., Parthasarathy, K.: Identification and control of dynamical systems using neural networks. IEEE Trans. Neural Netw. **1**(1), 4–27 (1990)
48. Zainuddin, Z., Pauline, O.: Modified wavelet neural network in function approximation and its application in prediction of time-series pollution data. Appl. Soft Comput. J. **11**(8), 4866–4874 (2011)
49. Langley, P., Simon, H.A.: Applications of machine learning and rule induction. Commun. ACM **38**(11), 54–64 (1995)
50. Larrañaga, P., Calvo, B., Santana, R., Bielza, C., Galdiano, J., Inza, I., Lozano, J.A., Armañanzas, R., Santafé, G., Pérez, A., Robles, V.: Machine learning in bioinformatics. Brief. Bioinf. **7**(1), 86–112 (2006)
51. Mjolsness, E., DeCoste, D.: Machine learning for science: state of the art and future prospects. Science **293**(5537), 2051–2055 (2001)
52. Ponce, H., Ponce, P., Molina, A.: The development of an artificial organic networks toolkit for labview. J. Comput. Chem. **36**(7), 478–492 (2015)
53. Srinivasan, K., Fisher, D.: Machine learning approaches to estimating software development effort. IEEE Trans. Softw. Eng. **21**(2), 126–137 (1995)
54. Zhu, Q., Qin, A.K., Suganthan, P.N., Huang, G.: Evolutionary extreme learning machine. Pattern Recogn. **38**(10), 1759–1763 (2005)
55. Geva, S., Sitte, J.: A constructive method for multivariate function approximation by multilayer perceptrons. IEEE Trans. Neural Netw. **3**(4), 621–624 (1992)
56. Ferrari, S., Stengel, R.: Smooth function approximation using neural networks. IEEE Trans. Neural Netw. **16**(1), 24–38 (2005). Cited By 108
57. Selmic, R., Lewis, F.: Neural-network approximation of piecewise continuous functions: Application to friction compensation. IEEE Trans. Neural Netw. **13**(3), 745–751 (2002). Cited By 89
58. Costarelli, D., Spigler, R.: Approximation by series of sigmoidal functions with applications to neural networks. Annali di Matematica Pura ed Applicata pp. 1–18 (2013). Cited By 4; Article in Press
59. Fateman, R.J.: Lookup tables, recurrences and complexity. In: Proceedings of the ACM-SIGSAM 1989 International Symposium on Symbolic and Algebraic Computation, ISSAC'89, pp. 68–73 (1989)
60. Liu, J., Liang, D.: A survey of FPGA-based hardware implementation of ANNs. In: ICNN B '05. International Conference on Neural Networks and Brain, 2005, vol. 2, pp. 915–918 (2005)
61. Mai-Duy, N., Tran-Cong, T.: Approximation of function and its derivatives using radial basis function networks. Appl. Math. Model. **27**(3), 197–220 (2003). Cited By 121
62. Misra, J., Saha, I.: Artificial neural networks in hardware: a survey of two decades of progress. Neurocomputing **74**(1–3), 239–255 (2010). Artificial Brains
63. Parri, J., Ratti, S.: Trigonometric function approximation neural network based coprocessor. In: Microsystems and Nanoelectronics Research Conference, 2009. MNRC 2009. 2nd, pp. 148–151 (2009)
64. Sahin, S., Becerikli, Y., Yazici, S.: Neural network implementation in hardware using FPGAS. In: King, I., Wang, J., Chan, L.W., Wang, D. (eds.) Neural Information Processing. Lecture Notes in Computer Science, vol. 4234, pp. 1105–1112. Springer, Berlin (2006)
65. Skoda, P., Lipic, T., Srp, A., Rogina, B., Skala, K., Vajda, F.: Implementation framework for artificial neural networks on fpga. In: Proceedings of the 34th International Convention MIPRO, 2011, pp. 274–278 (2011)
66. Dórea, C.E.T.: Set-invariant estimators for single-output linear discrete-time systems. Universidade Federal da Bahia, Tech. rep. (2008)

Adaptive Neuro-Fuzzy Controller of Induction Machine Drive with Nonlinear Friction

Abdesselem Boulkroune, Salim Issaouni and Hachemi Chekireb

Abstract In this chapter, a novel adaptive neuro-fuzzy backstepping control scheme is developed for induction machines with unknown model, uncertain load-torque and nonlinear friction. Neuro-fuzzy systems are used to online approximate the uncertain nonlinearities and an adaptive backstepping technique is employed to systematically construct the control law. The proposed adaptive fuzzy controller guarantees the tracking error converge to a small neighborhood of the origin and the boundedness of all closed-loop signals. These neuro-fuzzy systems are adjusted on-line according to some adaptation laws deduced from the stability analysis in the sense of Lyapunov. Compared to previous works, the proposed controller can effectively deal with the induction motors drives with both unified nonlinear frictions and (structured and unstructured) uncertainties. In fact, this present work can be seen as a non- trivial extension of the previous works. Simulation results are provided to demonstrate the effectiveness of the proposed control approach.

Keywords Induction motors · Adaptive neuro-fuzzy systems · Backstepping control · Nonlinear friction model

A. Boulkroune (✉)
LAJ, University of Jijel, BP. 98 Ouled-Aissa, 18000 Jijel, Algeria
e-mail: boulkroune2002@yahoo.fr

S. Issaouni · H. Chekireb
LCP, ENP, BP. 182 10 Avenue Hassan Badi El-Harrach, Algiers, Algeria
e-mail: autosalim@gmail.com

H. Chekireb
e-mail: chekireb@yahoo.fr

© Springer International Publishing Switzerland 2016   169
H.E. Ponce Espinosa (ed.), *Nature-Inspired Computing for Control Systems*,
Studies in Systems, Decision and Control 40, DOI 10.1007/978-3-319-26230-7_7

1 Introduction

The induction machine (IM) drives have been received a lot of attention in theoretical researches and practical applications. Since the machine currents are induced, no brushes and slip rings are required [21]. They are simple in operation, rugged and generally less expensive than either DC or synchronous motors. On the other hand, the control problem of IM is theoretically challenging, as such systems are multivariable, coupled, highly nonlinear and with uncertain model. Indeed, its model is much more complex than other machines, it is commonly used as "a benchmark problem in nonlinear systems". In addition, some uncertainties coming from load torque and rotor resistance are typically present in the model [21].

The field oriented control (FOC) can be seen a conventional control approach for IM. This control technique guarantees an efficient decoupling between torque and flux for IM and allows obtaining a DC machine similar model [21, 22]. Some advanced control techniques have been also used to tackle the speed and position control problems of IM, such as sliding mode control [23, 30], feedback linearization control [14, 27], adaptive backstepping control [1, 20, 29], and neural adaptive control [21]. However, these control approaches have some drawbacks, namely: (1) the friction model is assumed to be linear. (2) The unmatched uncertainties have not been considered.

In machine drives, the fuzzy logic control has been proposed as a promising alternative to conventional control techniques [13, 19]. Compared to conventional control, the principal advantage of this intelligent control approach resides in the fact that no mathematical model of the system is needed and the human experience can be systematically and easily incorporated in the control system via the fuzzy rules. Nevertheless, the non-adaptive fuzzy controllers are not robust to changing environment [13]. Therefore, it is crucial to add some learning forms that allows updating the control system parameters in order to keep and improve the control performance in a large range of varying conditions [13]. Using fuzzy logic systems for estimating of the uncertain functions, adaptive fuzzy control schemes for IM have been designed in [2, 15–17, 24, 33]. However, the control methods in [2, 15–17, 24, 33]. Suffer from some limitations, namely: (1) the friction model is not nonlinear and general. (2) The unmatched uncertainties have not been considered. (3) In [15–17] as a virtual control in a backstepping framework. This causes a problem during the stability analysis process.

Recently, in [13], a fuzzy adaptive controller based on backstepping approach has been designed for a class of doubly-fed induction motors. To effectively deal with the load charge, an online estimator has been constructed. In [18], an efficient adaptive tracking control approach with minimal learning parameters is designed for the 6-order induction motor model with a discrete-time form. A neuro-fuzzy adaptive control method for speed control of induction motor based on field oriented control has been constructed in [26]. In [32], a novel adaptive supervisory Takagi–Sugeno–Kang fuzzy cerebella model articulation control system in a sensorless vector control for induction motor drives has been proposed and online implemented. This controller can only control the rotor speed. In [28], a fuzzy adaptive sliding-mode controller based on the boundary layer approach for speed control of an indirect field-oriented control of induction motors has been proposed. This controller has been also tested in a real-time environment. However, the control methods in [18, 26, 28, 32] suffer from some limitations, namely: (1) the friction model is not nonlinear and general. (2) The unmatched uncertainties have not been considered. (3) The controller designed can only control the rotor speed.

Motivated by the above considerations, in this chapter, we will design a new adaptive neuro-fuzzy backstepping control system for IM drives which guarantees a flux regulation and an accurate speed tracking in the presence of both structured and unstructured uncertainties. Compared with the closely-related control methods [2, 15–18, 24, 26, 28, 32, 33], the main contributions of this chapter are:

- Unlike in [2, 15–18, 24, 26, 28, 32, 33], the frictions are assumed to be modeled by a nonlinear unknown unified model. Adopting such a model can provide a more accurate representation for the induction motor dynamics, but it brings hard nonlinearities to the system model.
- The resulting controller is not only robust to structured uncertainties (stator resistance, rotor resistance, etc.) and external disturbances (load torque), but also the system robustness can be achieved against the unstructured uncertainties (as unmodeled dynamics, nonlinear frictions,…).
- Unlike in [18, 26, 28, 32], the proposed controller can conjointly guarantee the control of both rotor flux and speed.
- The control system design does not depend on the IM's model.
- Unlike in [15–17], the stability analysis and control design are rigorously derived with mild assumptions.

2   Machine Modeling

In this section, we will present the nonlinear dynamical model of an IM. For the convenience of referencing, the meaning of the variables, parameters and coefficients is listed in Table 1 (in the Nomenclature).

Under the assumptions of equal mutual inductance and a linear magnetic circuit and via the field oriented transformation, the induction motor can be described, in the well-known (d-q) frame, by the following dynamic equations [27]:

$$
\begin{cases}
\frac{d\omega}{dt} = \frac{p\,L_m}{J\,L_r}\,\varphi_r\,i_{sq} - \frac{\tau_F}{J} - \frac{T_l}{J} \\
\frac{d\varphi_r}{dt} = \frac{L_m}{T_r}\,i_{sd} - \frac{1}{T_r}\,\varphi_r \\
\frac{d\,i_{sq}}{dt} = -\left(\frac{R_s}{\sigma\,L_s} + \frac{1-\sigma}{\sigma\,T_r}\right) i_{sq} - \omega_e\,i_{sd} - \frac{p\,L_m\,\omega}{\sigma\,L_s\,L_r}\,\varphi_r + \frac{1}{\sigma\,L_s}\,v_{sq} \\
\frac{d\,i_{sd}}{dt} = -\left(\frac{R_s}{\sigma\,L_s} + \frac{1-\sigma}{\sigma\,T_r}\right) i_{sd} + \omega_e\,i_{sq} - \frac{L_m}{\sigma\,L_s\,L_r\,T_r}\,\varphi_r + \frac{1}{\sigma\,L_s}\,v_{sd}
\end{cases}
\tag{1}
$$

Table 1 Nomenclature

R_s	Stator resistance	ω_e	Electric pulsation
R_r	Rotor resistance	ω	Angular mechanic velocity
L_s	Stator inductance	φ_{rd}	Rotor flux along d-axis
L_r	Rotor inductance	φ_{rq}	Rotor flux along q-axis
L_m	Mutual inductance	φ_r	Total rotor flux
p	Number of pole pairs	i_{sd}	Stator current along d-axis
J	Inertia	i_{sq}	Stator current along q-axis
T_l	Load torque	v_{sd}	Stator voltage along d-axis
σ	Blondel coefficient	v_{sq}	Stator voltage along q-axis
T_r	Rotor time constant	τ_F	Friction torque

$$\text{where,}\ \sigma = 1 - \frac{L_m^2}{L_s L_r},\ T_r = \frac{L_r}{R_r},\ \omega_e = \omega_d + p\omega,\ \omega_d = \frac{L_m}{T_r \varphi_r} i_{sq}.$$

The friction term τ_F is assumed to have the following non-linear parameterizable form [25]:

$$\tau_F = \lambda_1(\tanh(\lambda_2\,\omega) - \tanh(\lambda_3\,\omega)) + \lambda_4 \tanh(\lambda_5\,\omega) + \lambda_6\,\omega \qquad (2)$$

where, $\lambda_i \in \Re\ \forall i = 1,\ldots,6$ denote unknown positive constants.

It is worth noticing that the friction model in (2) is smooth, and includes all essential aspects of frictions [25]. This model has the following important features [25]:

- It is symmetric about the origin.
- The static coefficient of friction can be calculated by $\lambda_1 + \lambda_4$.
- The term $(\tanh(\lambda_2\,\omega) - \tanh(\lambda_3\,\omega))$ can present the Stribeck.
- The viscous dissipation term can be approximated by the term $\lambda_6\,\omega$.
- The coefficient of the Coulomb friction, being present when the viscous dissipation term is considered zero, can be presented by $\lambda_4 \tanh(\lambda_5\,\omega)$.

The model of the IM with external disturbances can be written as follows:

$$\begin{cases} \dot{x}_1 = f_1(x_1) + b_1(x_1)\,x_2 + d_1(x_1,x_2,t) \\ \dot{x}_2 = f_2(x_1,x_2) + b_2(x_1,x_2)u + d_2(x_1,x_2,u,t) \\ y = x_1 \end{cases} \qquad (3)$$

with

$$f_1(x_1) = \begin{bmatrix} -\frac{\tau_F}{J} \\ -\frac{1}{T_r}\varphi_r \end{bmatrix}, b_1(x_1) = \begin{bmatrix} \frac{p\,L_m}{JT_r}\varphi_r & 0 \\ 0 & \frac{L_m}{T_r} \end{bmatrix}, d_1(x_1,x_2,t) = \begin{bmatrix} -\frac{T_l}{J} + d_{11}(x_1,x_2,t) \\ d_{12}(x_1,x_2,t) \end{bmatrix}$$

$$f_2(x_1,x_2) = \begin{bmatrix} -\left(\frac{R_s}{\sigma L_s} + \frac{1-\sigma}{\sigma T_r}\right)i_{sq} - \omega_e\, i_{sd} - \frac{p\,L_m\,\omega}{\sigma L_s L_r}\varphi_r \\ -\left(\frac{R_s}{\sigma L_s} + \frac{1-\sigma}{\sigma T_r}\right)i_{sd} + \omega_e\, i_{sq} + \frac{L_m}{\sigma L_s L_r T_r}\varphi_r \end{bmatrix},$$

$$b_2(x_1,x_2) = \begin{bmatrix} \frac{1}{\sigma L_s} & 0 \\ 0 & \frac{1}{\sigma L_s} \end{bmatrix}, d_2(x_1,x_2,u,t) = \begin{bmatrix} d_{21}(x_1,x_2,u,t) \\ d_{22}(x_1,x_2,u,t) \end{bmatrix}$$

where $u = [v_{sq}, v_{sd}]^T$ is the control input of the system, $y = x_1 = [x_{11}, x_{12}]^T = [\omega, \varphi_r]^T$ is the controlled output, $x_2 = [x_{21}, x_{22}]^T = [i_{sq}, i_{sd}]^T$ is the vector of the stator currents, $x = [\omega, \varphi_r, i_{sq}, i_{sd}]^T$ represents the state vector. $f_i(.)$ and $b_i(.)$ are unknown smooth nonlinear functions, $d_i(.)$ is the term of unknown disturbances due to the parametric variations, non-modeled dynamics and external disturbances.

To facilitate the controller design, we rewrite the state model (3) in the following equivalent form:

$$\begin{cases} \dot{x}_1 = f_1(x_1) + g_1 x_2 + \delta_1(x_1,x_2,t) \\ \dot{x}_2 = f_2(x_1,x_2) + g_2 u + \delta_2(x_1,x_2,u,t) \\ y = x_1 \end{cases} \tag{4}$$

where

$$g_1 = \begin{bmatrix} \frac{p\,L_m}{JT_r} & 0 \\ 0 & \frac{L_m}{T_r} \end{bmatrix}, g_2 = b_2(x_1,x_2), \delta_2(x_1,x_2,u,t) = d_2(x_1,x_2,u,t)$$

and $\delta_1(x_1,x_2,t) = d_1(x_1,x_2,t) + (b_1(x_1) - g_1)x_2$. Note that g_1 and g_2 are unknown, constant and diagonal positive define matrices. Thereafter, for a better understanding, the following notations are made: $f_1 = f_1(x_1)$; $f_2 = f_2(x_1,x_2)$; $\delta_1 = \delta_1(x_1,x_2,t)$ and $\delta_2 = \delta_2(x_1,x_2,u,t)$.

Throughout this paper, one makes the following realistic assumptions.

Assumption 1 The reference speed profile ω_{ref} and the flux reference φ_{ref} are bounded and derivable.

Assumption 2 All motor parameters are assumed to be unknown.

Assumption 3 The load torque T_l is unknown but bounded.

Assumption 4 The nonlinear functions f_i and δ_i involved in the state-space model (4) are continuous and unknown.

Remark 1 The control design for the model (4) of the induction motor with matched and unmatched uncertainties and an unified nonlinear friction is very difficult and challenging. To authors' best knowledge, such a model with all these properties has never been considered in the previous works.

Remark 2 Assumptions 1–4 are mild and generally used in the literature.

3   Metaphor of the Neuro-Fuzzy System

The fuzzy logic system (FLS) and neural networks (NN) are two well-known metaphors. The FLS offer a mathematic calculus to interpret the subjective human knowledge of the real systems. In others words, these FLS can effectively represent intrinsic uncertainties of the human or expert knowledge with some linguistic variables. The artificial NN are a family of learning models inspired by biological neural networks (the biological functions of the human brain). The later is used to approximate (generally unknown) functions with a huge number of inputs.

In this chapter, the neuro-fuzzy system, which will be used to estimate the unknown nonlinear functions, is a fusion of these two metaphors. This resulting system offers the advantages of both NN and FLS and reduces their limitations. In fact, this system combines the learning power NN system and explicit knowledge representation of FLS.

Now, we will describe the neuro-fuzzy system (i.e. the zero-order Takagi-Sugeno (TS0) fuzzy system) used to approximate the unknown nonlinear functions being present in the induction motor's model. As shown in Fig. 1, there basically exist four parts for a neuro-fuzzy system, namely: the fuzzifier, the knowledge base (i.e. fuzzy rule base), the fuzzy inference engine and the defuzzifier.

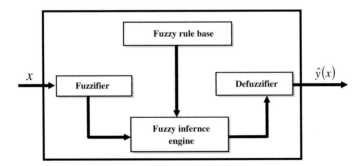

Fig. 1 Basic configuration of a fuzzy logic system

The fuzzy logic system performs a mapping for $U \subset \Re^4$ to $V \subset \Re$. Let $U = U_1 \times \ldots \times U_4$ where $U_i \subset \Re$, $i = 1, 2, 3, 4$. The fuzzy logic system is characterized by a set of IF-THEN rules in the following form:

$$R^{(l)} : \text{IF } x_1 \text{ is } F_1^l \text{ and} \ldots x_4 \text{ is } F_4^l \text{ THEN } y \text{ is } G^l \tag{5}$$

Such $l = 1, 2, \ldots, N$, N is the number of fuzzy rules for each the fuzzy model, and $x = [x_1, x_2, x_3, x_4] \in U$ and $y \in V$ are the input and output of the fuzzy systems respectively, F_i^l and G^l are fuzzy sets in U and V, respectively.

By using the singleton fuzzifier, product inference, and center-average defuzzifier, the output of the fuzzy system can be expressed as follows:

$$\hat{y}(x) = \theta^T \psi(x) \tag{6}$$

where $\theta^T = [\theta^1 \quad \theta^2 \quad \ldots \quad \theta^N] \in \Re^N$ is the adjustable parameters vector (i.e. consequent parameters), and $\psi^T = [\psi^1(x) \quad \psi^2(x) \ldots \quad \theta^N(x)] \in \Re^N$ is the fuzzy basis function (FBF) with $\psi^l(X)$ being expressed as:

$$\psi^l(x) = \prod_{i=1}^4 \mu_{F_i^l}(x_i) \Big/ \sum_{l=1}^N \left(\prod_{i=1}^4 \mu_{F_i^l}(x_i) \right) \tag{7}$$

$\mu_{F_i^l}(x_i)$ is the membership function of fuzzy set.

Notice that the neuro-fuzzy system (6) is usually used in control and identification applications [4–12, 31]. It has been proven in [31] that this neuro-fuzzy system (6) can estimate any continuous nonlinear function $y(x)$ over a compact operating space to an arbitrary degree of accuracy. Of particular importance, one assumes that the FBF vector, $\psi(x)$ is appropriately specified beforehand. But, The vector of the consequent parameters θ, will be online calculated via some appropriate update laws that will be designed later.

4 Design of Neuro-Fuzzy Adaptive Controller

The design of the fuzzy adaptive control by backstepping will be proceed in two stages:

Step 1: We note by $y_{ref} = [\omega_{ref}, \varphi_{ref}]^T$ the reference signal. In this case, the tracking error is given by:

$$e_1 = y_{ref} - y \tag{8}$$

Then, the tracking error dynamics are given by:

$$\dot{e}_1 = \dot{y}_{ref} - \dot{y} = \dot{y}_{ref} - f_1 - g_1 x_2 - \delta_1 \qquad (9)$$

The dynamics (9) can be rewritten as

$$g_1^{-1} \dot{e}_1 = F_1(x_1) - x_2 - g_1^{-1} \delta_1 \qquad (10)$$

where $F_1(x_1) = g_1^{-1}(\dot{y}_{ref} - f_1)$.

Since the function $F_1(x_1)$ is considered here to be unknown, we can use the neuro-fuzzy system (6) to approximate it as: $F_1(x_1) = \xi_1^T(x_1)\theta_1$ where $\xi_1^T(x_1)$ represents the regressive matrix and θ_1 is the adjustable parameter vector.

According to the fuzzy approximation property [3, 31], for given $\bar{\varepsilon}_1 > 0$, there exists a so-called optimal fuzzy system $\xi_1^T(x_1)\theta_1^*$ such that $F_1(x_1) = \xi_1^T(x_1)\theta_1^* + \varepsilon_1(x_1)$, where $\varepsilon_1(x_1)$ is the fuzzy approximation error and satisfies $|\varepsilon_1(x_1)| \le \bar{\varepsilon}_1$.

Then, from (10), the virtual control α can be constructed as

$$\alpha = \xi_1^T(x_1)\theta_1 + C_1 e_1 + v_{r1} \qquad (11)$$

where $C_1 = Diag[c_{11}, c_{12}]$ is a positive-define design matrix, $v_{r1} = [v_{r11}, v_{r12}]^T$ is a robust dynamic control term that will be designed later, and θ_1 is the estimate of the unknown constant θ_1^*.

Define a new state variable α_d. Let α pass through a first-order filter with a time constant κ to obtain α_d as:

$$\kappa \dot{\alpha}_d + \alpha_d = \alpha, \quad \text{where } \alpha_d(0) = \alpha(0) \qquad (12)$$

By defining $e_2 = \alpha_d - x_2$ and using (11) and (12), the dynamics (10) can be rewritten as follows:

$$\begin{aligned}
g_1^{-1} \dot{e}_1 &= \xi_1^T(x_1)\theta_1^* + \varepsilon_1(x_1) + e_2 + \Delta\alpha - \alpha - g_1^{-1}\delta_1 \\
&= \xi_1^T(x_1)\theta_1^* + \varepsilon_1(x_1) + e_2 + \Delta\alpha - \xi_1^T(x_1)\theta_1 - C_1 e_1 - v_{r1} - g_1^{-1}\delta_1 \quad (13) \\
&= -\xi_1^T(x_1)\tilde{\theta}_1 + e_2 - C_1 e_1 - v_{r1} + \Delta\alpha + \varepsilon_1(x_1) - g_1^{-1}\delta_1
\end{aligned}$$

where $\tilde{\theta}_1 = \theta_1 - \theta_1^*$ and $\Delta\alpha = \alpha - \alpha_d$.

Assumption 5 We assume that:

$$\left| \Delta\alpha + \varepsilon_1(x_1) - g_1^{-1}\delta_1 \right| \le (1 + \|x_1\| + \|\Delta\alpha\|) k_1^* \qquad (14)$$

where $k_1^* = [k_{11}^*, k_{12}^*]^T$ is an unknown constant vector.

The associated adaptive laws can be designed as

$$\dot{\theta}_1 = -\gamma_{\theta 1}\,\sigma_{\theta 1}\,\theta_1 + \gamma_{\theta 1}\,\xi_1(x_1)\,e_1 \tag{15}$$

$$\dot{v}_{r1} = -\gamma_{r1}\,v_{r1} + \gamma_{r1}\left[e_1 - \frac{v_{r1}}{v_{r1}^T v_{r1} + \delta_1^2}(1 + \|x_1\| + \|\Delta\alpha\|)\,k_1^T|e_1|\right] \tag{16}$$

$$\dot{\delta}_1 = -\gamma_{\delta 1}\,\sigma_{\delta 1}\,\delta_1 - \gamma_{\delta 1}\frac{\delta_1}{v_{r1}^T v_{r1} + \delta_1^2}(1 + \|x_1\| + \|\Delta\alpha\|)\,k_1^T|e_1| \tag{17}$$

$$\dot{k}_1 = -\gamma_{k1}\,\sigma_{k1}\,k_1 - \gamma_{k1}(1 + \|x_1\| + \|\Delta\alpha\|)|e_1| \tag{18}$$

where $\gamma_{\theta 1}, \gamma_{r1}, \gamma_{\delta 1}$ and γ_{k1} are strictly positive design constants, and $\sigma_{\theta 1}, \sigma_{\delta 1}$, and σ_{k1} are small positive design constants.

Consider the Lyapunov candidate function as follows:

$$V_1 = \frac{1}{2}e_1^T g_1^{-1} e_1 + \frac{1}{2\gamma_{\theta 1}}\tilde{\theta}_1^T \tilde{\theta}_1 + \frac{1}{2\gamma_{k1}}\tilde{k}_1^T \tilde{k}_1 + \frac{1}{2\gamma_{\delta 1}}\delta_1^2 + \frac{1}{2\gamma_{r1}}v_{r1}^T v_{r1} \tag{19}$$

where $\tilde{k}_1 = k_1 - k_1^*$.

Its time derivative is

$$\dot{V}_1 = e_1^T g_1^{-1} \dot{e}_1 + \frac{1}{\gamma_{\theta 1}}\tilde{\theta}_1^T \dot{\theta}_1 + \frac{1}{\gamma_{k1}}\tilde{k}_1^T \dot{k}_1 + \frac{1}{\gamma_{\delta 1}}\delta_1 \dot{\delta}_1 + \frac{1}{\gamma_{r1}}v_{r1}^T \dot{v}_{r1} \tag{20}$$

By considering (15)–(18), $\dot{V}_1$ can be bounded as

$$\dot{V}_1 \le -e_1^T C_1 e_1 + e_1^T e_2 - \sigma_{\theta 1}\tilde{\theta}_1^T \theta_1 - \sigma_{k1}\tilde{k}_1^T k_1 - \sigma_{\delta 1}\delta_1^2 - v_{r1}^T v_{r1} \tag{21}$$

In the next step, we will try to stabilize the tracking error e_2.

Step 2: At this step, we will construct the control law u. The dynamics of e_2 are given by

$$\dot{e}_2 = \dot{\alpha}_d - \dot{x}_2 = \dot{\alpha}_d - f_2 - g_2 u - \delta_2 \tag{22}$$

The dynamics (22) can be rewritten as follows

$$g_2^{-1}\dot{e}_2 = -e_1 + F_2(x) - u - g_2^{-1}\delta_2 \tag{23}$$

where $F_2(x) = g_2^{-1}(\dot{\alpha}_d - f_2) + e_1$.

Because $F_2(x)$ is considered here to be unknown, we can use the neuro-fuzzy system (6) to estimate it as: $F_2(x) = \xi_2^T(x)\,\theta_2$, where $\xi_2^T(x)$ represents the regressive matrix and θ_2 is the adjustable parameter vector.

According to the fuzzy approximation property [31], for given $\bar{\varepsilon}_2 > 0$, there exists a so-called optimal fuzzy system $\xi_2^T(x)\,\theta_2^*$ such that $F_2(x) = \xi_2^T(x)\,\theta_2^* + \varepsilon_2(x)$, where $\varepsilon_2(x)$ is the fuzzy approximation error and satisfies $|\varepsilon_2(x)| \leq \bar{\varepsilon}_2$.

To stabilize the dynamics (23), the following fuzzy adaptive controller is proposed

$$u = \xi_2^T(x)\,\theta_2 + C_2\,e_2 + v_{r2} \tag{24}$$

where $C_2 = Diag\,[c_{21}, c_{22}]$ is a positive-define design matrix, $v_{r2} = [v_{r21}, v_{r22}]^T$ is a robust dynamic control term that will be designed later, and θ_2 is the estimate of the unknown constant θ_2^*.

Using (24), the dynamics (23) can be rewritten as follows:

$$g_2^{-1}\,\dot{e}_2 = -e_1 - \xi_2^T(x)\,\tilde{\theta}_2 - C_2\,e_2 - v_{r2} + \varepsilon_2(x) - g_2^{-1}\,\delta_2 \tag{25}$$

where $\tilde{\theta}_2 = \theta_2 - \theta_2^*$.

Assumption 6 We assume that:

$$\left|\varepsilon_2(x) - g_2^{-1}\,\delta_2\right| \leq (1 + \|x\| + \|u\|)\,k_2^* \tag{26}$$

where $k_2^* = \left[k_{21}^*, k_{22}^*\right]^T$ is an unknown constant vector.

The associated adaptive laws can be designed as

$$\dot{\theta}_2 = -\gamma_{\theta2}\,\sigma_{\theta2}\,\theta_2 + \gamma_{\theta2}\,\xi_2(x)\,e_2 \tag{27}$$

$$\dot{v}_{r2} = -\gamma_{r2}\,v_{r2} + \gamma_{r2}\left[e_2 - \frac{v_{r2}}{v_{r2}^T\,v_{r2} + \delta_2^2}(1 + \|x\| + \|u\|)\,k_2^T|e_2|\right] \tag{28}$$

$$\dot{\delta}_2 = -\gamma_{\delta2}\,\sigma_{\delta2}\,\delta_2 - \gamma_{\delta2}\frac{\delta_2}{v_{r2}^T\,v_{r2} + \delta_2^2}(1 + \|x\| + \|u\|)\,k_2^T|e_2| \tag{29}$$

$$\dot{k}_2 = -\gamma_{k2}\,\sigma_{k2}\,k_2 - \gamma_{k2}(1 + \|x\| + \|u\|)|e_2| \tag{30}$$

where $\gamma_{\theta2}, \gamma_{r2}, \gamma_{\delta2}$ and γ_{k2} are strictly positive design constants, and $\sigma_{\theta2}, \sigma_{\delta2}$ and σ_{k2} are small positive design constants.

Define the Lyapunov function candidate as follows:

$$V_2 = V_1 + \frac{1}{2}e_2^T\,g_2^{-1}\,e_2 + \frac{1}{2\gamma_{\theta2}}\tilde{\theta}_2^T\,\tilde{\theta}_2 + \frac{1}{2\gamma_{k2}}\tilde{k}_2^T\,\tilde{k}_2 + \frac{1}{2\gamma_{\delta2}}\delta_2^2 + \frac{1}{2\gamma_{r2}}v_{r2}^T\,v_{r2} \tag{31}$$

Its time derivative is

$$\dot{V}_2 = \dot{V}_1 + e_2^T g_2^{-1} \dot{e}_2 + \frac{1}{\gamma_{\theta 2}} \tilde{\theta}_2^T \dot{\theta}_2 + \frac{1}{\gamma_{k2}} \tilde{k}_2^T \dot{k}_2 + \frac{1}{\gamma_{\delta 2}} \delta_2 \dot{\delta}_2 + \frac{1}{\gamma_{r2}} v_{r2}^T \dot{v}_{r2} \qquad (32)$$

By considering (27)–(30), $\dot{V}_2$ can be bounded as

$$
\begin{aligned}
\dot{V}_2 \leq & -e_2^T C_2 e_2 - e_1^T C_1 e_1 - \sigma_{\theta 1} \tilde{\theta}_1^T \theta_1 - \sigma_{k1} \tilde{k}_1^T k_1 - \sigma_{\delta 1} \delta_1^2 - v_{r1}^T v_{r1} \\
& - \sigma_{\theta 2} \tilde{\theta}_2^T \theta_2 - \sigma_{k2} \tilde{k}_2^T k_2 - \sigma_{\delta 2} \delta_2^2 - v_{r2}^T v_{r2}
\end{aligned} \qquad (33)
$$

We can show that:

$$
\begin{aligned}
-\sigma_{\theta i} \tilde{\theta}_i^T \theta_i &\leq -\frac{\sigma_{\theta i}}{2} \left\| \tilde{\theta}_i \right\|^2 + \frac{\sigma_{\theta i}}{2} \left\| \tilde{\theta}_i^* \right\|^2 \\
-\sigma_{ki} \tilde{k}_i^T k_i &\leq -\frac{\sigma_{ki}}{2} \left\| \tilde{k}_i \right\|^2 + \frac{\sigma_{ki}}{2} \left\| \tilde{k}_i^* \right\|^2
\end{aligned} \qquad (34)
$$

Using (34), $\dot{V}_2$ becomes:

$$
\begin{aligned}
\dot{V}_2 \leq & -c_{1m} \|e_1\|^2 - c_{2m} \|e_2\|^2 - \frac{\sigma_{\theta 1}}{2} \left\| \tilde{\theta}_1 \right\|^2 - \frac{\sigma_{\theta 2}}{2} \left\| \tilde{\theta}_2 \right\|^2 - \frac{\sigma_{k1}}{2} \left\| \tilde{k}_1 \right\|^2 - \frac{\sigma_{k2}}{2} \left\| \tilde{k}_2 \right\|^2 \\
& - \sigma_{\delta 1} \delta_1^2 - v_{r1}^T v_{r1} - \sigma_{\delta 2} \delta_2^2 - v_{r2}^T v_{r2} + \pi
\end{aligned} \qquad (35)
$$

where $\pi = \frac{\sigma_{\theta 1}}{2} \left\| \theta_1^* \right\|^2 + \frac{\sigma_{\theta 2}}{2} \left\| \theta_2^* \right\|^2 + \frac{\sigma_{k1}}{2} \left\| k_1^* \right\|^2 + \frac{\sigma_{k2}}{2} \left\| k_2^* \right\|^2$, $c_{1m} = \min\{c_{11}, c_{12}\}$ and $c_{2m} = \min\{c_{21}, c_{22}\}$.

From (31) and (35), we get:

$$\dot{V}_2 \leq -\mu V_2 + \pi \qquad (36)$$

where

$$\mu = \min \left\{ \frac{2 c_{1m}}{\mu_{\max}\left(g_1^{-1}\right)}, \frac{2 c_{2m}}{\mu_{\max}\left(g_2^{-1}\right)}, \gamma_{\theta 1} \sigma_{\theta 1}, \gamma_{\theta 2} \sigma_{\theta 2}, \gamma_{k1} \sigma_{k1}, \gamma_{k2} \sigma_{k2}, 2 \gamma_{\delta 1} \sigma_{\delta 1}, 2 \gamma_{\delta 2} \sigma_{\delta 2}, 2 \gamma_{r1}, 2 \gamma_{r2} \right\}$$

Integrating (36) over $[0, t]$ yields

$$0 \leq V_2(t) \leq \frac{\pi}{\mu} + \left[V_2(0) - \frac{\pi}{\mu} \right] e^{-\mu t} \qquad (37)$$

From (37), it is clear that the tracking errors and the parametric estimation errors exponentially converge to a bounded adjustable domain. Figure 2 illustrates the diagram of the proposed fuzzy backstepping controller of the IM.

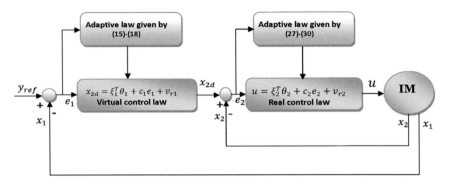

Fig. 2 Illustrative diagram of the adaptive fuzzy backstepping controller of the IM

Remark 3 Compared to the previous works [15–18, 24, 26, 28, 32], the neuro-fuzzy control approach described in this paper has the following important advantages:

- Compared to previous closely related works [15–18, 26, 28, 32], the flux regulation and accurate speed tracking are conjointly assured by a smooth fuzzy controller given by (11) and (24). In other words, our proposed controller is free of the undesirable chattering generally due to the use the signum function in the control law [15–17, 26, 28, 32].
- The system robustness can be achieved against matched and unmatched uncertainties of the induction motor. In other words, our control system can be effectively deal with a general form of uncertainties being present in all stages of the induction machine model.
- Unlike in [15–18, 24, 26, 28, 32], the frictions are assumed here to be modeled by a nonlinear unknown unified model. The later can capture all the most essential aspects of frictions [25].
- The control design does not strongly depend on the induction motor model, i.e. our controller is not applicable for a particular model of induction motor.
- Unlike in [15–17], the stability analysis and control design are rigorously derived in mathematics. Note that in [15–17], some non-derivable control terms have been used as a virtual control in a backstepping framework. This leads therefore to a non-rigorous proof of the stability analysis.

Remark.4 Neuro-fuzzy logic systems used in this chapter represents a class of linearly parameterized approximators, and can be replaced by any linearly parameterized approximator without any additional effort, such as Radial Basis Function (RBF) networks, and wavelet networks. However, only FLS provide a systematic framework to integrate the linguistic fuzzy information provided from human experts.

5 Conventional Backstepping Design

In this subsection, we will design a conventional backstepping controller. To facilitate the controller design, we rewrite the model (1) in the following equivalent form:

$$
\begin{cases}
\dot{x}_1 = a_1 x_2 x_3 - D \\
\dot{x}_2 = -\gamma x_2 - \omega_e x_4 - b_3 x_1 x_3 + c v_{sq} \\
\dot{x}_3 = a_2 x_4 - b_2 x_3 \\
\dot{x}_4 = -\gamma x_4 + \omega_e x_2 - b_4 x_3 + c v_{sd} \\
y = x_1
\end{cases} \tag{38}
$$

where $x_1 = \omega$, $x_2 = i_{sq}$, $x_3 = \varphi_r$ and $x_4 = i_{sd}$,

$$
a_1 = \frac{p\,L_m}{J\,L_r}, \quad a_2 = \frac{L_m}{T_r}, \quad D = \frac{\tau_F + T_l}{J}, \quad \gamma = \frac{R_s}{\sigma\,L_s} + \frac{1-\sigma}{\sigma\,T_{r\,2}},
$$

$$
b_2 = \frac{1}{T_r}, b_3 = \frac{p\,L_m}{\sigma\,L_s\,L_r}, b_4 = \frac{L_m}{\sigma\,L_s\,L_r\,T_r}, c = \frac{1}{\sigma\,L_s}.
$$

The design procedure of this conventional backstepping control scheme contains four steps.

Step 1: For the reference signal x_{1ref}, we define the tracking error as $e_1 = x_{1ref} - x_1$. Therefore, from (38), the error dynamics can be given by $\dot{e}_1 = \dot{x}_{1ref} - \dot{x}_1 = \dot{x}_{1ref} - a_1 x_2 x_3 + D$.

Choosing a Lyapunov function candidate as $V_1 = \frac{1}{2} e_1^2$, then its time derivative is

$$
\dot{V}_1 = e_1 \dot{e}_1 = e_1 \left(\dot{x}_{1ref} - a_1 x_2 x_3 + D \right) \tag{39}
$$

Design the virtual control law α_1 as

$$
\alpha_1 = \frac{1}{a_1 x_3} \left(K_1 e_1 + D + \dot{x}_{1ref} \right) \tag{40}
$$

with $K_1 > 0$ is a design constant.

Define $e_2 = \alpha_1 - x_2$, the expression of $\dot{e}_1$ can be rewritten as

$$
\dot{e}_1 = -K_1 e_1 + a_1 x_3 e_2 \tag{41}
$$

Using (41), obviously, the time derivative of V_1 becomes

$$
\dot{V}_1 = -K_1 e_1^2 + a_1 e_1 e_2 x_3 \tag{42}
$$

In the next step, we will try to stabilize the tracking error e_2.

Step 2: From (38), dynamics of e_2 can be given by

$$\dot{e}_2 = \dot{\alpha}_1 - \dot{x}_2 = \dot{\alpha}_1 + \gamma x_2 + \omega_e x_4 + b_3 x_1 x_3 - c v_{sq} \tag{43}$$

Now, we consider a Lyapunov function candidate as $V_2 = V_1 + \frac{1}{2} e_2^2$. Therefore, its time derivative is

$$\dot{V}_2 = \dot{V}_1 + e_2 \dot{e}_2 = -K_1 e_1^2 + e_2 \left(\dot{\alpha}_1 + a_1 e_1 x_3 + \gamma x_2 + \omega_e x_4 + b_3 x_1 x_3 - c v_{sq} \right) \tag{44}$$

And the control input v_{sq} can be constructed as

$$v_{sq} = \frac{1}{c} \left(K_2 e_2 + \dot{\alpha}_1 + a_1 e_1 x_3 + \gamma x_2 + \omega_e x_4 + b_3 x_1 x_3 \right) \tag{45}$$

where $K_2 > 0$ is a design constant.

Furthermore, using (45), we get

$$\dot{e}_2 = -K_2 e_2 \tag{46}$$

and

$$\dot{V}_2 = -K_1 e_1^2 - K_2 e_2^2 \tag{47}$$

Step 3: For a reference signal x_{3ref}, we define the tracking error as $e_3 = x_{3ref} - x_3$. From (38), the dynamics of the error e_3 is $\dot{e}_3 = \dot{x}_{3ref} - \dot{x}_3 = \dot{x}_{3ref} - a_2 x_4 + b_2 x_3$.

Choosing the Lyapunov function candidate as $V_3 = V_2 + \frac{1}{2} e_3^2$. Then, its time derivative is

$$\dot{V}_3 = \dot{V}_2 + e_3 \dot{e}_3 = -K_1 e_1^2 - K_2 e_2^2 + e_3 \left(\dot{x}_{3ref} - a_2 x_4 + b_2 x_3 \right) \tag{48}$$

From (48), we can design the virtual control law α_2 as follows:

$$\alpha_2 = \frac{1}{a_2} \left(K_3 e_3 + b_2 x_3 + \dot{x}_{3ref} \right) \tag{49}$$

with $K_3 > 0$ is a design constant.

Let's define $e_4 = \alpha_2 - x_4$, then we can express the dynamics of e_3 as

$$\dot{e}_3 = -K_3 e_3 + a_2 e_4 \tag{50}$$

Using (49), (48) becomes

$$\dot{V}_3 = -K_1 e_1^2 - K_2 e_2^2 - K_3 e_3^2 + a_2 e_3 e_4 \tag{51}$$

In the last step, we will stabilize the tracking error e_4.

Step 4: Now, we will design the control law v_{sd}. To this end, we consider a Lyapunov function candidate as $V_4 = V_3 + \frac{1}{2} e_4^2$. Its time derivative is

$$\dot{V}_4 = \dot{V}_3 + e_4 \dot{e}_4 = -\sum_{i=1}^{3} K_i e_i^2 + e_4(\dot{\alpha}_2 + a_2 e_3 + \gamma x_4 - \omega_e x_2 + b_4 x_3 - c v_{sd})$$

$$(52)$$

with

$$\dot{\alpha}_2 = \frac{1}{a_2} \left(-(K_3 - b_2) \dot{x}_3 + K_3 \dot{x}_{3ref} + \ddot{x}_{3ref} \right) \tag{53}$$

From (52), we can design the control law as

$$v_{sq} = \frac{1}{c} (K_4 e_4 + \dot{\alpha}_2 + a_2 e_3 + \gamma x_4 - \omega_e x_2 + b_4 x_3) \tag{54}$$

with $K_4 > 0$ is a design constant.

Furthermore, using (54), we obtain

$$\dot{e}_4 = -K_4 e_4 \tag{55}$$

$$\dot{V}_4 = -\sum_{i=1}^{4} K_i e_i^2 \leq 0 \tag{56}$$

From (56), it is obvious that the closed loop system is stable and all tracking errors converge asymptotically to zero.

6 Simulation Results

The parameter values of the IM are given in Table 2. This proposed control system has been implemented in a MATLAB-SIMULINK environment. The reference of speed is selected as follows:

$$\omega_{ref} = 60 + 60 \sin(\pi t) \; [\text{rd/s}]$$

The reference of rotor flux is $\varphi_{ref} = 1 \, \text{Wb}$. It is supposed that the induction machine is subjected to a load of 5 N m at 0.6 s. The parameters of the friction model are selected as: $\lambda_1 = 15$, $\lambda_2 = 200$, $\lambda_3 = 10$, $\lambda_4 = 13$, $\lambda_5 = 200$ and $\lambda_6 = 0.014$.

The fuzzy system $\xi_1^T(x_1)\theta_1$ has the vector $x_1 = [x_{11}, x_{12}]^T$ as input, while the fuzzy system $\xi_2^T(x_1, x_2)\theta_2$ has the state vector $x = [x_1^T, x_2^T]^T = [x_{11}, x_{12}, x_{21}, x_{22}]^T$ as input. For each variable of the entries of these fuzzy systems, as in Fig. 3, we define

Table 2 Parameters of the induction motor

Parameter	Value	Parameter	Value
Nominal power	1.5 KW	Nominal speed	1420 tr/min
Nominal current	3.8/6.5 A	Nominal voltage	220 V/380 V
Stator resistance (R_s)	4.85 Ω	Nominal frequency	50 Hz
Rotor resistance (R_r)	3.805 Ω	Number of pair of poles (p)	2
Stator inductance (L_s)	0.274 H	Blondel coefficient (σ)	0.113
Rotor inductance (L_r)	0.274 H	Rotor time constant (T_r)	0.072 s
Mutual inductance (L_m)	0.258 H	Inertia (J)	0.031 Kg m^2

three (one triangular and two trapezoidal) membership functions uniformly distributed on the following intervals: $[0\ 140]$ for x_{11}, $[0\ 1.4]$ for x_{12}, and $[-8\ 8]$ for x_{21} and x_{22}.

After numerous tests, we have selected the design parameters of the fuzzy adaptive controller giving a satisfactory performance, as: $\gamma_{\theta 1} = 20$, $\gamma_{\theta 1} = 30$, $\sigma_{\theta 1} = 20$, $\sigma_{\theta 2} = 30$, $C_1 = diag[40,\ 900]$, $C_2 = diag[40,\ 40]$, $\kappa = 0.02$, $\gamma_{r1} = 25$, $\gamma_{\delta 1} = 10^{-6}$, $\sigma_{\delta 1} = 10^{-7}$, $\gamma_{k1} = 10^{-5}$, $\sigma_{k1} = 3$, $\gamma_{r2} = 25$, $\gamma_{\delta 2} = 10^{-6}$, $\sigma_{\delta 2} = 10^{-7}$, $\gamma_{k2} = 10^{-5}$, $\sigma_{k2} = 3$.

In order to overcome the singularity problem, the initial conditions of adaptation laws v_{rj} and δ_j are selected as: $v_{r1}(0) = v_{r2}(0) = [2,2]^T$, $\delta_1(0) = \delta_2(0) = 2$. For other adaptive parameters, the initial conditions are arbitrarily selected as: $k_1(0) = k_2(0) = [10,\ 10]^T$ and $\theta_{1j}(0) = \theta_{2j}(0) = 0.01$ for $j = 1, \ldots, m$ where m is the number of the fuzzy rules.

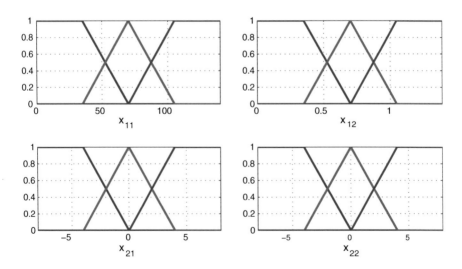

Fig. 3 Fuzzification of the fuzzy system inputs x_{ij}

After many tests, the parameters of the conventional backstepping controller are adopted as follows: $K_1 = 1500$, $K_2 = 300$, $K_3 = 40$, $K_4 = 300$.

For both controllers, in the simulation experimental, four cases are considered:

Case 1   Simulation without nonlinear frictions but with nominal parameters of the induction motor
Case 2   Simulation with nonlinear frictions and nominal parameters
Case 3   Simulation with nonlinear frictions and parametric variations (R_s is amplified to 200 %)
Case 4   Simulation with nonlinear frictions and parametric variations (L_s and L_r decreased by 30 %)

The simulation results of both controllers without considering nonlinear frictions and parametric variations are shown respectively in Figs. 4 and 5. According to these results, we notice that rotor flux respect the constraints of orientation and the speed follows perfectly its imposed reference.

The Figs. 6, 7, 8, 9, 10 and 11 show the performance of both controllers with friction model and parametric variations (i.e. the stator resistance R_s is amplified to

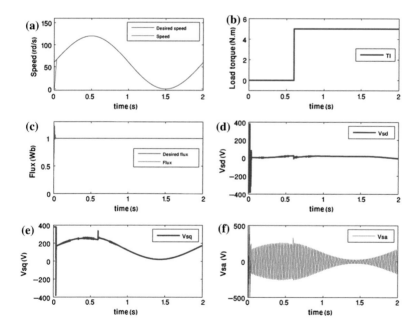

Fig. 4 Response of IM for adaptive fuzzy backstepping control without nonlinear frictions and with nominal parameters: **a** The speed and its reference signal; **b** Load torque; **c** The flux and its reference signal; **d** Stator voltage along d-axis; **e** Stator voltage along q-axis; **f** First component of the 3ph stator voltage (Vsa)

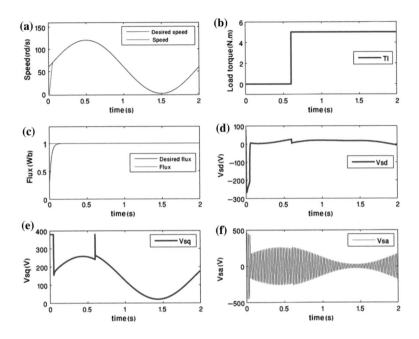

Fig. 5 Response of IM for conventional backstepping control without nonlinear frictions and with nominal parameters: **a** The speed and its reference signal; **b** Load torque; **c** The flux and its reference signal; **d** Stator voltage along d-axis; **e** Stator voltage along q-axis; **f** First component of the 3ph stator voltage (Vsa)

200 % and L_s and L_r decreased by 30 %). From these figures, we can see that the adaptive fuzzy backstepping control presents a good performance and is robust to parametric variations, compared with the conventional backstepping controller.

At time 1.5 s, the motor is stopped. When it starts again, we can see that the adaptive fuzzy backstepping controller follows quickly and perfectly his reference, unlike to the conventional backstepping controller.

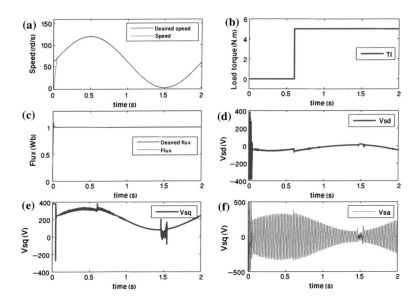

Fig. 6 Response of IM for adaptive fuzzy backstepping control with nonlinear frictions and with nominal parameters: **a** The speed and its reference signal; **b** Load torque; **c** The flux and its reference signal; **d** Stator voltage along d-axis; **e** Stator voltage along q-axis; **f** First component of the 3ph stator voltage (Vsa)

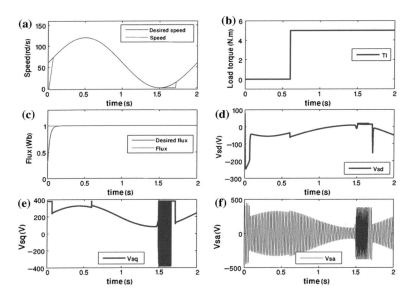

Fig. 7 Response of IM for conventional backstepping control with nonlinear frictions and with nominal parameters: **a** The speed and its reference signal; **b** Load torque; **c** The flux and its reference signal; **d** Stator voltage along d-axis; **e** Stator voltage along q-axis; **f** First component of the 3ph stator voltage (Vsa)

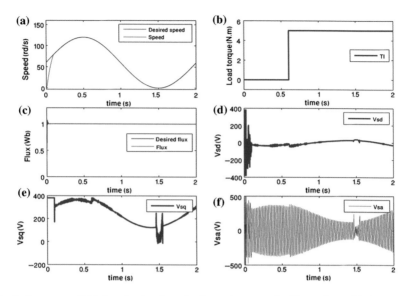

Fig. 8 Response of IM for adaptive fuzzy backstepping control with nonlinear frictions and R_s amplified to 200 %: **a** The speed and its reference signal; **b** Load torque; **c** The flux and its reference signal; **d** Stator voltage along d-axis; **e** Stator voltage along q-axis; **f** First component of the 3ph stator voltage (Vsa)

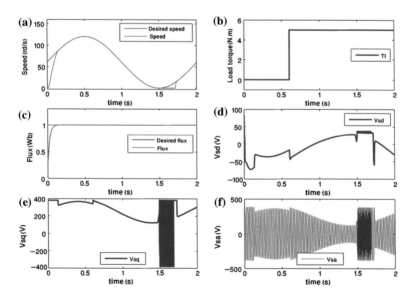

Fig. 9 Response of IM for conventional backstepping control with nonlinear frictions and R_s amplified to 200 %: **a** The speed and its reference signal; **b** Load torque; **c** The flux and its reference signal; **d** Stator voltage along d-axis; **e** Stator voltage along q-axis; **f** First component of the 3ph stator voltage (Vsa)

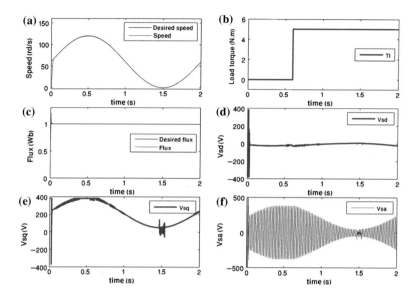

Fig. 10 Response of IM for adaptive fuzzy backstepping control with nonlinear frictions and L_s and L_r decreased by 30 %: **a** The speed and its reference signal; **b** Load torque; **c** The flux and its reference signal; **d** Stator voltage along d-axis; **e** Stator voltage along q-axis; **f** First component of the 3ph stator voltage (Vsa)

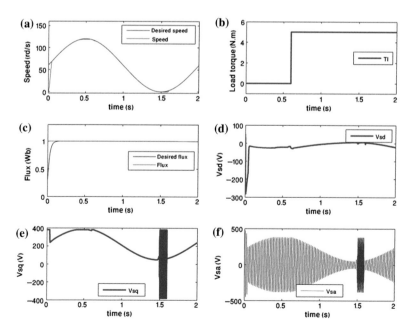

Fig. 11 Response of IM for conventional backstepping control with nonlinear frictions and L_s and L_r decreased by 30 %: **a** The speed and its reference signal; **b** Load torque; **c** The flux and its reference signal; **d** Stator voltage along d-axis; **e** Stator voltage along q-axis; **f** First component of the 3ph stator voltage (Vsa)

Table 3 Tracking performance comparison between both controllers

		Without τ_F	With nonlinear frictions τ_F		
			The nominal case	$2R_s$	$0.7L_s$ and $0.7L_r$
Adaptive fuzzy backstepping	MSE speed	26.66	34.76	56.84	20.94
	MSE flux	9.91×10^{-4}	8.91×10^{-3}	10^{-3}	7.96×10^{-4}
Conventional backstepping	MSE speed	41.66	65.18	105.75	36.95
	MSE flux	4×10^{-3}	4×10^{-3}	4×10^{-3}	4×10^{-3}

MSE mean square error

Based on the above simulation results, tracking performance for both cases of the proposed adaptive fuzzy backstepping control and conventional backstepping control in presence of nonlinear frictions and parametric variations is summarized in Table 3 which confirm that the control is robust to parametric variations.

7 Conclusion

In this chapter, an adaptive fuzzy backstepping controller of IM with unknown model, uncertain load-torque and nonlinear frictions has been presented. Fuzzy systems have been used to online estimate uncertain functions and an adaptive backstepping technique has been employed to systematically construct the control law. The corresponding adaptive laws have been derived in Lyapunov sense to guarantee a practical stability.

Compared with the existing control methods on fuzzy adaptive control of induction motors, the main advantages of this contribution are: (1) The frictions are assumed to be modeled by a nonlinear unified model thereby providing a more accurate representation for the induction motor dynamics. (2) The proposed controller is not only robust to structured uncertainties but also to unstructured uncertainties. However, the main drawbacks of this contribution is the need of a huge number of design parameters. Of particular interest, in a realistic numerical simulation framework, the proposed neuro-fuzzy control system has shown to be more efficient in performing the tracking control than the conventional backstepping control.

It is worth noting that the control approach proposed here can be effortlessly extended to any other high performance electric drives. In the future work, one will focus on the experimental validation of this proposed controller and the design of sensorless control schemes.

References

1. Arbin, E., Gregory, M.: Adaptive backstepping control of a speed-sensorless induction motor under time-varying load torque and rotor resistance uncertainty. In: Proceedings of the 38th Southeastern Symposium on System Theory, Tennessee Technological University Cookeville, TN, USA, pp. 512–518, March 5–7 (2006)
2. Agamy, M., Youcef, H.A., Sebakhy, O.A.: Adaptive fuzzy variable structure control of induction motors. In: Proceedings of the Canadian Conference of Computer and Electrical Engineering, pp. 89–94, Niagara Falls, Canada (2004)
3. Boulkroune, A., Tadjine, M., M'saad, M., Farza, M.: How to design a fuzzy adaptive control based on observers for uncertain affine nonlinear systems. Fuzzy Sets Syst. **159**, 926–948 (2008)
4. Boulkroune, A., M'saad, M.: A fuzzy adaptive variable-structure control scheme for uncertain chaotic MIMO systems with sector nonlinearities and dead-zones. Expert Syst. Appl. **38**, 4744–4750 (2011)
5. Boulkroune, A.: M'saad, M.: On the design of observer-based fuzzy adaptive controller for nonlinear systems with unknown control gain sign. Fuzzy Sets Syst. **201**, 71–85 (2012)
6. Boulkroune, A., M'saad, M.: Fuzzy adaptive observer-based projective synchronization for nonlinear systems with input nonlinearity. J. Vib. Control, **18**, 437–450 (2012)
7. Boulkroune, A., M'saad, M., Farza, M.: Adaptive fuzzy tracking control for a class of MIMO nonaffine uncertain systems. Neurocomptiung, **93**, 48–55 (2012)
8. Boulkroune, A., M'saad, M., Farza, M.: Fuzzy approximation-based indirect adaptive controller for multi-input multi-output non-affine systems with unknown control direction. IET Control Theory Appl. **17**, 2619–2629 (2012)
9. Boulkroune, A., Bounar, N., M'saad, M., Farza, M.: Indirect adaptive fuzzy control scheme based on observer for nonlinear systems: A novel SPR-filter approach. Neurocomputing, 135: 378–387 (2014)
10. Boulkroune, A., M'saad, M., Farza, M.: State and output feedback fuzzy variable structure controllers for multivariable nonlinear systems subject to input nonlinearities. Int. J. Adv. Manuf. Technol. **71**, 539–556 (2014)
11. Boulkroune, A., Tadjine, M., M'Saad, M., Farza, M.: Design of a unified adaptive fuzzy observer for uncertain nonlinear systems. Inf. Sci. **265**, 139–153 (2014)
12. Boulkroune, A., Bouzeriba, A., Hamel, S., Bouden, T.: A projective synchronization scheme based on fuzzy adaptive control for unknown multivariable chaotic systems. Nonlinear Dyn. **78**(1), 433–447 (2014)
13. Bounar, N., Boulkroune, A., Boudjema, F.: Fuzzy adaptive controller for a DFI-motor. In: Complex System Modelling and Control Through Intelligent Soft Computations (pp. 87–110). Springer International Publishing, New York City (2015)
14. Chiasson, J.: A new approach to dynamic feedback linearization control of an induction motor. IEEE Trans. Autom. Control **43**, 391–397 (1998)
15. Ezziani, N., Essounbouli, N., Hamzaoui, A.: Backstepping fuzzy adaptive controller of induction machine. In Proceedings of the 16th Mediterranean Conference on Control and Automation, Ajaccio, France, pp. 1622–1627 (2008)
16. Ezziani, N., Essounbouli, N., Hamzaoui, A.: Backstepping adaptive type-2 fuzzy controller for induction machine. In: IEEE international symposium on Indus. Electronics, Cambridge (UK), pp. 443–448 (2008)
17. Ezziani, N., Essounbouli, N., Hamzaoui, A.: An AFB controller of induction machine under amplitude and rate saturation constraints. In: 35th Annual Conference of IEEE Industrial Electronics, 2009. IECON'09, pp. 1026–1032 (2009)
18. Gao, Y., Wang, H., Liu, Y.J.: Adaptive fuzzy control with minimal leaning parameters for electric induction motors. Neurocomputing **156**, 143–150 (2015)
19. Ghamri, A., Benchouia, M.T., Benbouzid, M.E.H., Golea, A., Zouzou, S.E.: Simulation and control of AC/DC converter and induction machine speed using adaptive fuzzy controller. In:

Proceedings of International Conference on Electrical Machines and Systems, Seoul, Korea, pp. 539–542 (2007)

20. Hwang, Y.H., Park, K.K., Yang, H.W.: Robust adaptive backstepping control for efficiency optimization of induction motors with uncertainties. In: IEEE International Symposium on Industrial Electronics, pp. 878–883 (2008)
21. Kwan, C.M., Lewis, F.L.: Robust backstepping control of induction motors using neural networks. IEEE Trans. Neural Netw. **11**(5), 1178–1187 (2000)
22. Krause, P.C.: Analysis of Electric Machinery. McGraw-Hill, New York, NY (2000)
23. Liu, J.K., Sun, F.C.: Research and development on theory and algorithms of sliding mode control. Control Theory Appl. **24**(3), 407–418 (2007)
24. Lin, F.J., Shen, P.H., Hsu, S.P.: Adaptive sliding mode control for linear induction motor drive, IEEE Proc. Power Appl. **149**(3), 184–194 (2002)
25. Makkar, C., Dixon, W.E., Sawyer, W.G., Hu, G.: A new continuously differentiable friction model for control systems design. In: Proceedings of the 2005 IEEE/ASME, International Conference on Advanced Intelligent Mechatronics, Monterey, California, USA (2005)
26. Masumpoor, S., Khanesar, M.A.: Adaptive sliding-mode type-2 neuro-fuzzy control of an induction motor. Expert Syst. Appl. **42**(19), 6635–6647 (2015)
27. Marino, R., Peresada, S., Valigi, P.: Adaptive input-output linearizing control of induction motors. IEEE Trans. Autom. Control **38**(2), 208–221 (1993)
28. Saghafinia, A., Ping, H.W., Uddin, M.N., Gaeid, K.S.: Adaptive fuzzy sliding-mode control into chattering-free IM drive. IEEE Trans. Ind. Appl. **51**(1), 692–701 (2015)
29. Tan, Y., Chang, J., Tan, H.: Adaptive backstepping control and friction compensation for AC servo with inertia and load uncertainties. IEEE Trans. Ind. Electron. **50**(5), 944–952 (2003)
30. Wai, R.J., Lin, K.M., Lin, C.Y.: Total sliding-mode speed control of field oriented induction motor servo drive. In: Proceedings of the 5th Asian Control Conference, Australia, pp. 1354–1361 (2004)
31. Wang, L.X.: Adaptive Fuzzy Systems and Control: Design and Stability Analysis. Prentice-Hall, Englewood Cliffs, NJ (1994)
32. Wang, S.Y., Tseng, C.L., Chiu, C.J.: Online speed controller scheme using adaptive supervisory TSK-fuzzy CMAC for vector controlled induction motor drive. Asian J. Control **16**(5), 1–13 (2015)
33. Youcef, H.A., Wahba, M.A.: Adaptive fuzzy MIMO control of induction motors. Expert Syst. Appl. **36**(1), 4171–4175 (2009)

Part III
Robotics Applications

Fuzzy Logic Sugeno Controller Type-2 For Quadrotors Based on Anfis

Pedro Ponce, Arturo Molina, Israel Cayetano, Jose Gallardo,
Hugo Salcedo, Jose Rodriguez and Isela Carrera

Abstract Artificial intelligence has opened new alternatives to control non-linear systems. One of the most important methods is the fuzzy logic controller, which is constructed with linguistic rules; however, it is normally based on the knowledge from human experts, so the linguistic rules and membership functions are not optimized. In the case of Quadrotors, it is important to acquire the knowledge from human experts because they are able to naturally describe the controller for this non-liner system by linguistic rules in a complete form but the memory space and processing time have to be minimum in the hardware implementation. Hence, a neuro fuzzy controller (ANFIS) is used in order to reduce the number of linguistic rules and membership functions and to preserve the surface between inputs and outputs in the controller for Quadrotors. Hence, the real time controller improves its response. This proposal keeps the basic idea of getting the knowledge from human experts and then ANFIS can be implemented in the real time hardware in the Quadrotor. The results show that this methodology is an excellent option to control Quadrotors.

1 Introduction

The interest of many investigators in the fields of UAVs (Unmanned Aerial Vehicles) has increased in the last years, primary in Defense, Government and commercial sectors. The ability to remotely control a vehicle capable to realize dangerous tasks without compromising the integrity of a human being is a big attraction to their use. Applications such as military operations, national security, border patrol, surveillance and combat are hazardous subjects in which UAVs result useful [9].

P. Ponce (✉) · A. Molina · I. Cayetano · J. Gallardo · H. Salcedo · J. Rodriguez · I. Carrera
Tecnologico de Monterrey, Campus Ciudad de México, Mexico, Mexico
e-mail: pedro.ponce@itesm.mx

© Springer International Publishing Switzerland 2016
H.E. Ponce Espinosa (ed.), *Nature-Inspired Computing for Control Systems*,
Studies in Systems, Decision and Control 40, DOI 10.1007/978-3-319-26230-7_8

One of the most used UAV that have had a boom in recent years due to many new applications is the Quadrotor (or Quadcopter), a multi-rotor helicopter. It is a machine with a crossed form where in the extreme of each arm is mounted a motor with its respective attached propeller. This configuration supposes a sum of single moving parts. Four smaller motors provide minimal mechanical complexity, payload augmentation and less damage in case of accidents. Besides, there is a reduction of gyroscopic effects and no need of gearing, [4] along with simplicity in design and construction [11]. As counterpart of its simplicity, the quadcopters flight requires a complex and precise control to handle the four motors interacting simultaneously [20, 26], a fundamental aspect that made their implementation initially unviable due to the low development of sensing and computing technology. Since the development of fast and high demand computing devices in the last decade, a large amount of research of quadrotors have been seen in many developed countries, mostly in the fields of control and design. Table 1, taken from Bouabdallah, summarizes some quadrotor related investigations until 2007 [3]. Mesicopter project consisted in a centimeter scale quadrotors, contrary to the P. Pounds thesis, which is based on the X-4 designed by the Australian National University with measurements of 1 m length and a 4 kg weight [24].

Table 1 Quadrotor works in various universities

Project	University	Status	Description
Mesicopter	Stanford	Finished	It is a cm-scale rotorcraft Application with low cost devices
E. **Table.** thesis	Univ. Pennsylvania	Finished	Using visual feedback as the primary sensor Two types of controller: Series of mode-based feedback linearizing controllers and Backstepping-like control law Tilt-roll rotor quadrotor
P. Castillo thesis	Univ. Compiegne	Finished	Predictor-observer based discrete-time controller
P. Pounds thesis	ANU	Finished	Develop a rotor design technique for maximizing thrust and the application of a novel rotor mast configuration
N. Guenard's thesis	CEA	Finished	Vision-based navigation strategy for a Vertical Take-off and Landing (VTOL) Unmanned Aerial Vehicle (UAV), using a single embedded camera observing natural landmarks
STARMAC	Stanford	In progress	Multi agent control of quadrotos of about 1 kg
M. Kemper's thesis	Univ. Oldenburg	Finished	2D GPS-based position control system for 4 Rotor Helicopters
G.P. Tournier thesis	MIT	Finished	Vision-based estimation of the position and orientation of an object using a single camera Six degree of freedom estimation Utilizing geometry and 4 single-point discrete Fourier transforms

The STARMAC project uses the Dragonflyer III and aimed an autonomous way-point tracker flight control system, and the creation of a multi-vehicle platform for experimentation and validation of multi-agent control algorithms [11]. The other studies showed in the Table 1 can be found in [1, 5, 10, 16, 30].

The study of Fuzzy Logic controllers for Quadrotors has been developed highly in the last decade. Coza and Macnab presented in 2006 a new Robust Adaptive-Fuzzy control method for Quadrotor stabilization. They propose a method using control and center updates for each axis rotation to approximate the same nonlinear function as the e-modification method. Results on Simulink showed that this new method offered a better center tuning than the e-method, obtaining less error in the steady state although oscillations were present in a range of ±0.2 radians [6, 7].

In 2010 Santos et al. [28] developed a PID-like type-1 fuzzy controller for the three axes of a Quadrotor and the altitude using trapezoid membership functions. The controller was tested in Simulink showing a smooth, fast, stable response [28]. The same year, Kirly et al. presented their work of the design of a Fuzzy controller embedded into a TMS320F28335 micro processing unit, testing the axes separately. They obtained that starting from 3 to 12°, the controller was able to reach its steady state of ±2° near-horizontal and good response to perturbations. They note also that given the sensitiveness of their inclination sensor, vibrations given by environmental noises are to be studied extensively [17].

In 2013 Sheikpour and Shouraki published their results of the design of a Fuzzy controller using the Parallel Distributed Compensation method obtained from a Takagi-Sugeno fuzzy model of a Quadrotor. They show the viability of this method obtaining a ~ 1 s response with a ~ 0.1 radians of overshoot [29]. The same year, Ilhan [13] presented their work in Type-2 Fuzzy Logic controller for a Quadrotor for position and altitude using triangular membership functions. It showed a very slightly better response than type-1 Fuzzy approach and significantly better than PID. However, it was not delved in the tuning and selection of different Footprints of Uncertainty [13].

In 2014, it was presented a Real Time Fuzzy controller embedded in a GUMSTIX Overo FIRESTORM COM microcontroller board. The controller was obtained using the ANFIS system from test data obtained from a first experiment. The results showed that fuzzy controller is easily capable of controlling the Quadrotor, with the advantage that it was self-tuned as opposed to the PD controller. Besides, Fuzzy outperformed PD in certain conditions [2]. Finally in 2014, a Hybrid method of backstepping and fuzzy adaptive PID is proposed by Qingji, Fengfa and Dandan [25]. In it, a Fuzzy Inference System is used to tune the parameters Kp, Ki and Kd of a PID controller. Simulation and practical results showed that this hybrid controller performed achieved the stabilization, effectiveness and robustness desired, with variations of ±1° in the steady state and rejection of disturbances of $\sim 5°$ [25], Table 2 summarize different controllers and advantages.

Table 2 Quadrotors controller and advantages

Controller	Advantages
GPS-base position control	Able to keep positions above given destinations as well as to navigate between waypoints while minimizing trajectory errors Enables permanent full speed flight with reliable altitude keeping considering that the resulting lift is decreasing while changing pitch or roll angles for position control
Vision based control	Allows the navigation of the vehicle when GPS lost the signal, it does not need to recover the position of the UAV with respect to a reference frame
Mode-based-feedback linearizing controllers	Switch between many modes such as hover, takeoff, landing, left/right, search, tilt-up, tilt-down etc.
Backstepping-like control law	Useful when some states are controlled through other states
Adaptive robust sliding mode control	Stability and robustness against unknown uncertainty and disturbances
Nonlinear control	Able to carry out the tasks of taking off, hovering, and positioning

1.1 Quadrotor Basic Principles

It is important to describe how the Quadrotor works in order to design its controller. The arrangement of motors in the Quadrotor is shown in Fig. 32. Each rotor has a thrust and an angular momentum about its center of rotation, as well as a drag force opposite to the rotorcraft's direction of flight [4]. To produce lift, the rotors have to spin at a certain speed to produce enough thrust. The quantity of thrust will determine the altitude and speed at which the Quadrotor rises [12]. On the other hand, the spinning of each motor generates an angular momentum that will try to rotate the Quadrotor in its yaw angle such as it would happen in a helicopter without tail propeller. To avoid this effect, two intercalated rotors will spin clockwise and the other two counterclockwise so that the individual angular momentum cancel each other. Therefore, the propellers attached to each motor are different: two of them are pusher and two are puller, working in contra-rotation.

In order to change the angle along pitch and roll axes, and therefore move it in a certain direction, depending on the desired speed of the displacement, the rotors orientated toward that direction must change their thrust as seen in the Fig. 1, but always taking care that the sum of angular momentum remains zero so that the Quadrotor remains in the same altitude and yaw angle.

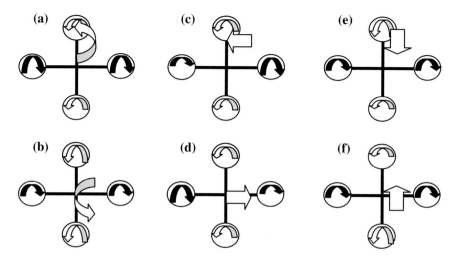

Fig. 1 Quadrotor frame and functioning (**a** clockwise yaw, **b** anticlockwise yaw, **c** anticlockwise roll, **d** clockwise roll, **e** anticlockwise pitch, **f** clockwise pitch)

2 Quadrotor Model

The Model of Rigid Body Dynamics of the airframe is exposed in [19] and explained next: Two reference fames are defined to find the position and orientation at it is showed at Fig. 2.

{A} is the inertial reference frame.
{B} is the body fixed frame.

The orientation of the rigid body is given by a rotation matrix in (1)

$$^{A}R_B = \begin{bmatrix} c\psi c\theta - s\phi s\psi s\theta & -c\phi s\psi & c\psi s\theta + c\theta s\phi s\psi \\ c\theta s\psi + c\psi s\phi s\theta & c\phi c\psi & s\psi s\theta - c\psi c\theta s\phi \\ -c\phi s\theta & s\phi & c\phi c\theta \end{bmatrix} \qquad (1)$$

where c and s mean sine and cosine. The rotation is model by XYZ Euler angles to get from A to B, with rotation angles ψ yaw angle, φ rotation about the x axis in the intermediary frame {E}, followed by third pitch rotation angle θ. The rigid body equation of motion are (2)–(4).

$$\dot{\xi} = v_p \qquad (2)$$

$$m\dot{v} = mg\vec{a}_3 + RF \qquad (3)$$

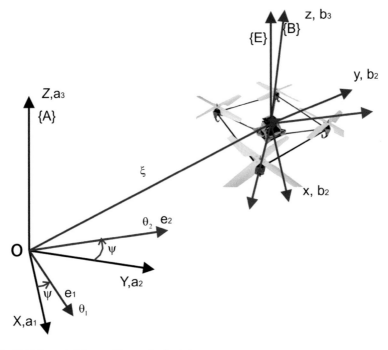

Fig. 2 Vehicle model for position and orientation

$$\dot{R} = R\Omega \times \text{ where } \mathbf{I}\Omega = -\Omega \times \mathbf{I}\Omega + \tau \qquad (4)$$

v denotes the linear velocity of {B} respect to frame{A}
Ω is the angular velocity of {B} respect to {A} expressed in {B}
m denotes the mass of the rigid body
I is the Inertia mass
Ωx is the skew—matrix
And F and m are principal non-conservative forces and moments applied to the quadrotor airframe by the aerodynamics

The aerodynamic model is useful for the design of rotor systems, where the whole range of parameters (rotor geometry, profile, hinge mechanism, and much more) are fundamental to the design problem. A basic level of aerodynamic modeling is required.

In steady state thrust generated by hovering rotor can be modeled by the simplified momentum theory when cT > 0, it is modeled as a constant that can be easily determined from static thrust tests Eq. (5).

$$T_i = c_T \varpi_i^2 \qquad (5)$$

The reaction torque (due to rotor drag) acting on the airframe generated by a hovering rotor in free air may be modeled as (6)

$$Q_i = c_Q \varpi_i^2 \qquad (6)$$

where the coefficient c_Q can be determinate by static thrust test.

The total thrust at hover (T_Σ) in (7) applied to the airframe is the sum of the thrusts from each individual rotor:

$$T_\Sigma = \sum_{i=1}^{N} |T_i| = c_T \left(\sum_{i=1}^{N} \varpi_i^2 \right) \qquad (7)$$

The hover thrust is the primary component of the exogenous force (8)

$$F = T_\Sigma \vec{z} + \Delta \qquad (8)$$

where Δ comprises secondary aerodynamic forces that are induced when the assumption that the rotor is in hover is violated.

The net moment arising from the aerodynamics written in Matrix form (9) (the combination of the produced rotor forces and air resistances)

$$\begin{pmatrix} T_\Sigma \\ \tau_1 \\ \tau_2 \\ \tau_3 \end{pmatrix} = \underbrace{\begin{pmatrix} c_T & c_T & c_T & c_T \\ 0 & dc_T & 0 & -dc_T \\ -dc_T & 0 & dc_T & 0 \\ -c_Q & c_Q & -c_Q & c_Q \end{pmatrix}}_{\Gamma} \begin{pmatrix} \varpi_1^2 \\ \varpi_2^2 \\ \varpi_3^2 \\ \varpi_4^2 \end{pmatrix} \qquad (9)$$

The rotor speed can be calculated using the inverse of constant matrix Γ. Suitable elements must be chosen in order for the vehicle to hover, such $\tau = 0$ and $T\Sigma = mg$.

2.1 Blade Flapping

There are many aerodynamic and gyroscopic effects associated with any rotorcraft that modify the simple force model introduced above. Blade flapping (showed in Fig. 3) and induced drag, however, are fundamental effects that are of significant importance in understanding the natural stability of quadrotors and how state observers operate. These effects are particularly relevant since they induce forces in the x-y rotor plane of the quadrotor, the under actuated directions in the dynamics, that cannot be easily dominated by high gain control.

Fig. 3 Flapping angle

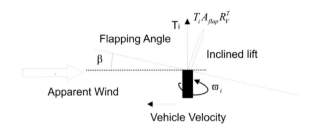

Blade-flapping effects are due to the flexing of rotors, while induced drag is associated primarily with the rigidity of the rotor, and a typical quadrotor will experience both.

When a rotor translates laterally through the air it displays an effect known as rotor flapping. A detailed derivation of rotor flapping involves a mechanical model of the bending of the rotor subject to aerodynamic and centripetal forces as it is swept through a full rotation [18].

The resulting equations of motion are a nonlinear second-order dynamical system with a dominant highly damped oscillatory response at the forced frequency corresponding to the angular velocity of the rotor. For a typical rotor, the flapping dynamics converge to steady state with one cycle of the rotor [18], and for the purposes of modeling, only the steady-state response of the flapping dynamics need be considered.

The flapping angle (10), β is the steady-state tilt of the rotor away from the incoming apparent wind and $\beta^{\perp}$ is the tilt orthogonal to the incident wind.

$$
\begin{aligned}
\beta &:= -\frac{\mu A_{1c}}{\left(1 - \frac{1}{2}\mu^2\right)}, \\
\beta^{\perp} &:= -\frac{\mu A_{1s}}{\left(1 - \frac{1}{2}\mu^2\right)}
\end{aligned}
\tag{10}
$$

μ is the advance radio,

$\mu := \frac{|V_x|}{\varpi r}$ radio of magnitude of horizontal velocity of the rotor to the linear velocity of the rotor tip.

For a general movement of the vehicle, results in the apparent wind forward ratio (11)

$$
\mu = \sqrt{v_x'^2 + v_y'^2} / \varpi r
\tag{11}
$$

$$
A_{flap} = \frac{1}{\varpi R}
\begin{pmatrix}
A_{1c} & -A_{1s} & 0 \\
A_{1s} & A_{1c} & 0 \\
0 & 0 & 0
\end{pmatrix}
\tag{12}
$$

This matrix (12) describes the sensitivities of the flapping angle to the apparent wind in the body-fixed frame. The first row encodes (9) for the velocity along the body fixed frame x axis. The second row of A_{flap} is a $\Pi/2$ rotation of this response to account for the case where a component of the wind is incoming from the y axis, while the third row projects out velocity in the z axis of the body fixed frame. In the body-fixed frame the induced drag is (13)

$$D_{ind.}v' \approx diag(d_x, d_y, 0)v' \tag{13}$$

where $dx = dy$ the induced drag coefficient.

The exogenous force (14) applied to the rotor can now be modeled by

$$F := T_\Sigma \vec{z} - T_\Sigma Dv' \tag{14}$$

where $D = Aflap + diag(dx, dy)$ y T_Σ is the nominal thrust be the projection matrix (15) onto the x-y plane.

$$\mathbb{P}_h := \begin{pmatrix} 1 & 0 & 0 \\ 0 & 1 & 0 \end{pmatrix} \tag{15}$$

The horizontal component of a velocity expressed in $\{A\}$ is (16)

$$v_h := \mathbb{P}_h v = (v_x, v_y)^T \in \mathbb{R}^2 \tag{16}$$

Projecting onto the horizontal component of velocity (17),

$$m\dot{v}_h = -T_\Sigma \mathbb{P}_h(\vec{z} + RDv') \tag{17}$$

If the vehicle is flying horizontally, i.e., $v_z = 0$, then $v = \mathbb{P}_h^T v_h$ and one can write Eq. (18)

$$m\dot{v}_h = -T_\Sigma \mathbb{P}_h \overset{r}{z} - \mathbb{P}_h RDR^T \mathbb{P}_h^T v_h \tag{18}$$

where the last term introduces damping since, for a typical system, the matrix D is a positive semi-definite.

2.2 Basic Diagram for Fuzzy Logic Type 2 Using in the Quadrotor

Figure 4 shows the main block diagram for a Fuzzy Logic Controller Type-2 (T2FS) which is similar to a traditional Fuzzy Logic Type 1 presented in this chapter.

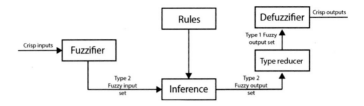

Fig. 4 Block diagram of a type 2 fuzzy logic system

Fuzzifier: The fuzzifier maps the read value into the T2DS which activates the inference system. Rule base: The antecedents and consequents are related each other by IF-ELSE language functions.

Inference: This block assign to the fuzzy input a fuzzy output according to the rules established and operators such as union ($\sqcup$) and intersection operators ($\sqcap$); those operators are equivalent to the union and intersection operations but are used in the secondary membership functions. They definition and explanation can be found in the paper presented by Mendel, 2006 [21].

Type reduction: For some systems it is required to transform the type 2 fuzzy outputs from the inference engine into T1FS and the result is called a type reduced set. There most common method for doing this are the Mendel 2007 [22] iteration algorithm and the uncertainty bounds method [21]. Both are based on the calculation of centroid of mass. Defuzzification: Once the outputs have been reduced the defuzzification block determines the crisp value that will be send to the actuator [21].

3 Fuzzy Logic

3.1 Type-1 Fuzzy Logic Control

Fuzzy Logic was introduced by the mathematician Lofti Zadeh, taking as example the way people think in order to make decisions: humans process in the form of linguistic variables the information they acquire from the environment. A significant advantage of Fuzzy Logic is that, as an extension of the classic Boolean logic, it allows to assign to a variable a degree of membership of a given characteristic which is in the interval [0,1]; that means that a variable cannot be only completely true or completely false, but also partially [27].

The Imperial College of Science's professor Ebrahim Mamdani proposed the use of Fuzzy Logic into a Control technique. The implementation of a control system based on Fuzzy Logic is divided in three steps. During the first step, the fuzzification, the crisp input, which is a numeric value, has to be translated into linguistic variables with its respective degree of membership. This is done through the use of Membership Functions, which depending on the input value can assign a value of $0°$ of membership in the case of a total absence of membership and 1 when there is

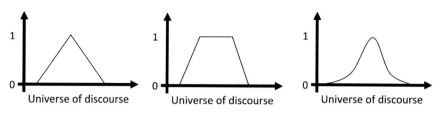

Fig. 5 Triangular, Trapezoidal and Gaussian forms of MFs

a total membership. It is necessary to make the Membership Functions cover all the domain of inputs or, in other words, Universe of Discourse. The most common shapes of this Membership Functions (MF) are trapezoidal, triangular, Gaussian, singleton among others (Fig. 5). Defining the shape and number of MFs will depend on the nature of the system, but mostly in the experience of the designer about the model to control, relying on what the designer finds best suits to the handling of input data [8].

The second step in the Fuzzy control is the inference. It involves the relationship among the input membership functions and the output control action to be applied. That connection is given by the use of inference rules with logical operators. For a Single Input Single Output system (a system with only one input X and one output Y) a commonly used structure of these rules is the following:
If X is A then Y is C.

For a system with multiple variables the rules can be done with logical operations as follows:
If X is A and Y is B then Z is C.

That means that membership functions A and B of the inputs X and Y are related to the membership function of the output Z. Following the principles of logical operations, it is considered that this type of assignation rule is applicable only when the operands A and B are TRUE; however, in fuzzy logic they will apply just if they are NOT FALSE; in other words, the degree of membership of each function should have any value within the interval [0,1]. Provided that both values are different than zero, it is possible to assign a degree of membership for the output of the rule, which could be taken as the multiplication of the degree of membership of the input functions involved (19).

$$\mu_{C_n} = \mu_A \mu_B \tag{19}$$

There are times when the output has more than one MF. In that case, the result of the inference is the union of each MF, creating a new combined MF as shown in Fig. 6.

The final step in the Fuzzy control is named defuzzification, a process in which the different linguistic values obtained for the output during the application of the rules are translated into a physical value, in other words, into a crisp numeric value. This step has the same principle of the fuzzification, but in defuzzification the value

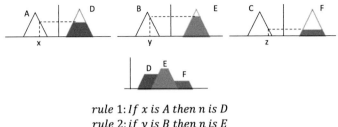

rule 1: *If x is A then n is D*
rule 2: *if y is B then n is E*
rule 3: *if z is C then n is F*

Fig. 6 Example of how a new combined MF is created

is converted from a degree of membership into a numeric value defined by the universe of discourse of the output. If different rules generate a value for uCn, it can be taken only the maximum value because it provides the greater weight to the membership function according to the minimax criterion. Defined that value, and obtained the combined MF, the most common way to calculate the crisp output is through the calculation of the center of mass of that MF, using the formula (20):

$$Z_0 = \frac{\mu_{C1}M_1 + \mu_{C2}M_2 + \cdots + \mu_{Cn}M_n}{\mu_{C1} + \mu_{C2} + \cdots + \mu_{Cn}} \qquad (20)$$

where Z_0 is the crisp output value, u_{Cn} is the degree of membership of each output membership function and M_n its respective output value.

3.2 Type-2 Fuzzy Control

Sometimes it is difficult to determine an exact membership function for a fuzzy set, because it depends on what the controller's designers consider follows the linguistic variable associated to it. This can happen when the input data has an uncertainty in the measurement due to noise in the signal, or if the knowledge used to make rules in a Fuzzy logic system is uncertain; therefore, this leads to have uncertain antecedents and/or consequences, translated into membership functions.

In order to reduce the effect of these uncertainties, Jerry Mendel proposed the use of type-2 Fuzzy Logic, the general case of type-1 Fuzzy Logic, to design controllers focusing on the mean and variance as their probabilistic modeling for random uncertainty. The defuzzified output of a type-1 Fuzzy Logic System (FLS) is analog to computing the mean of a probability density function; as variance provides a measure of dispersion about the mean, FLS's also need measure of dispersion to capture more rule uncertainties than just a single number. Type-2 Fuzzy Logic provides the dispersion fundamental to design systems with linguistic and numerical uncertainties in the degree of membership, using sets that handle linguistic uncertainties, a necessity because "a word can mean different things to different people". Figure 7 shows the form of a type-2 Fuzzy Set. It is characterized by its 2D Footprint of Uncertainty (FOU) bound by a lower and upper Membership Functions (LMF and UMF) and its embedded type-1 Fuzzy Sets [23, 27, 32].

Figure 8 shows the structure of a type-2 FLS. It can be seen from the Fig. 8 that, unlike type-1, type-2 FLS includes a block called Type Reducer before the defuzzification step. The fuzzifier maps the crisp input into a fuzzy set. The rules stays the same as type-1 FLS, "R: if x1 is F1 and x2 is F2 and… and xp Is Fp, THEN y is G". It is not necessary that all the antecedents and consequents must be type-2 fuzzy sets; it is only needed one antecedent or consequent type-2 set to make a type-2 FLS.

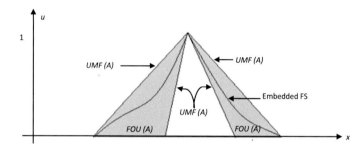

Fig. 7 Example of a type-2 MF

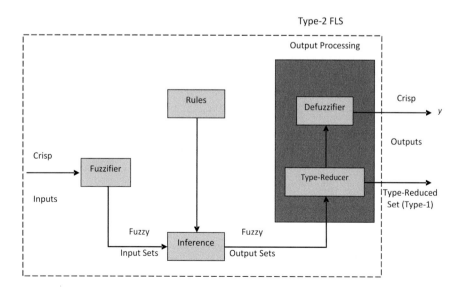

Fig. 8 Scheme of the implemented type 2 fuzzy logic controller

The inference in type-2 FLS combines rules and gives a mapping form input type-2 fuzzy sets to output type-2 fuzzy sets, as well as compositions of type-2 relations. As the output of the inference engine is a type-2 set, the defuzzifier uses "extended versions" of type-1 defuzzification methods resulting to a type-1 fuzzy set. This operation is called type reduction. Next, an explanation of inference and type reduction is detailed.

3.3 *Inference in Type-2 FLS*

We use Fig. 9 to explain how the inference in Type-2 FLS is made. There, when $x_1 = x'_1$, a vertical line intersects FOU $F_1^\%$ everywhere in the interval $\left[\mu_{F_1^\%}(x'_1), \overline{\mu_{F_1^\%}}(x'_1)\right]$. An analogue case happens when $x_2 = x'_2$ (now in the interval $\left[\mu_{F_2^\%}(x'_2), \overline{\mu_{F_2^\%}}(x'_2)\right]$). Then, two firing levels are computed, a lower $\underline{f}(x')$ and a upper firing levels $\overline{f}(x')$ being (21)–(23).

$$\underline{f}(x') = min\left[\underline{\mu_{F_1^\%}}(x'_1), \underline{\mu_{F_2^\%}}(x'_2)\right] \tag{21}$$

And

$$\overline{f}(x') = min\left[\overline{\mu_{F_1^\%}}(x'_1), \overline{\mu_{F_2^\%}}(x'_2)\right] \tag{22}$$

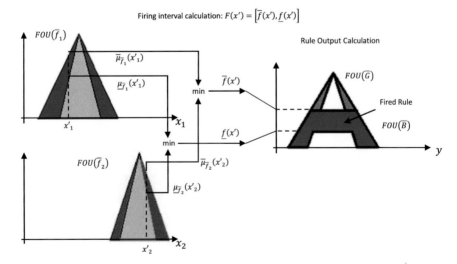

Fig. 9 Explanation of the inference calculation

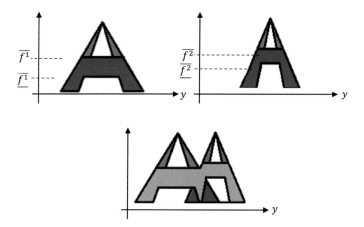

Fig. 10 Combined type-2 fuzzy set resulted from the inference

$$\bar{f}(x') = min\left[\overline{\mu_{F_1^{\%}}}\left(x_1'\right), \overline{\mu_{F_2^{\%}}}\left(x_2'\right)\right] \tag{23}$$

The result is a firing interval F(x'), where $F(x')\left[\underline{f}(x'), \bar{f}(x')\right]$, with $\underline{f}(x')$ the t-normed with LMF ($G^{\%}$) and $\bar{f}(x')$ the t-normed with UMF ($G^{\%}$). The union of the induced outputs for each rule in the inference section creates a final output from which the crisp output will be obtained. Figure 10 is an example for this.

Once the inference is done (getting a type-2 Fuzzy Set), a type reduction method is needed in order to make the defuzzification. This method is an extension of type-1 defuzzification method based on a centroid calculation getting a Reduced-Type Set.

3.4 Finding the Crisp Output in T2 FLS

For an induced type-2 Fuzzy Set $B^{\%}$, a centroid $C^{\%}$ is defined by $\left[y_l\left(B^{\%}\right), y_r\left(B^{\%}\right)\right]$ being y_l the smallest value and y_r the largest. The Karnik-Mendel iterative Algorithm [15] is used to obtain the centroid $C^{\%}$. Figure 11 show a type-2 Fuzzy Set in order to get a better idea of what these formulas are calculating. c_l (24) and c_r (25) are the end points of the centroids of the consequent Induced type-2 FSs.

$$y_l = y_l(L) = \frac{\sum_{i=1}^{L} y_i UMF\left(\widetilde{B}|y_i\right) + \sum_{i=L+1}^{N} y_i LMF\left(\widetilde{B}|y_i\right)}{\sum_{i=1}^{L} UMF\left(\widetilde{B}|y_i\right) + \sum_{i=L+1}^{N} LMF\left(\widetilde{B}|y_i\right)} \tag{24}$$

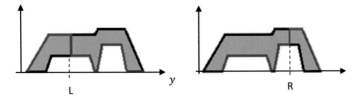

Fig. 11 Karnik-Mendel calculation of the center of mass, [15] and a fast method for computing thecentroid of a type-2 fuzzy set is presented in [31]

$$y_r = y_r(R) = \frac{\sum_{i=1}^{R} y_i UMF\left(\widetilde{B}|y_i\right) + \sum_{i=R+1}^{N} y_i LMF\left(\widetilde{B}|y_i\right)}{\sum_{i=1}^{R} UMF\left(\widetilde{B}|y_i\right) + \sum_{i=R+1}^{N} LMF\left(\widetilde{B}|y_i\right)} \tag{25}$$

Once calculated, the crisp output will be their average (26):

$$y(x) \approx \widehat{y}(x) = \frac{1}{2}[\widehat{y}_l(x) + \widehat{y}_r(x)] \tag{26}$$

Another method for obtaining the crisp output to the type-2 FLS is called the Nie-Tan. In this, the centroid, and therefore the output, of the inferred type-2 Fuzzy Set is obtained by the formula:

$$y_{output} = \frac{\sum_{i=1}^{N} \left(\overline{\mu}_{inferred}(y_i) + \underline{\mu}_{inferred}(y_i)\right) y_i}{\sum_{i=1}^{N} \overline{\mu}_{inferred}(y_i) + \underline{\mu}_{inferred}(y_i)} \tag{27}$$

It can be seen that this formula is very similar to the KM method, but its calculation returns directly the crisp output.

4 ANFIS

Fuzzy logic and neural networks are complementary tools for building intelligent systems. While neural networks are low-level computational structures that perform well when dealing with raw data, fuzzy logic deals with reasoning on a higher level, using linguistic information acquired from domain experts. However, fuzzy systems lack the ability to learn and cannot adjust themselves to a new environment. On the other hand, although neural networks can learn, they are black boxes to the user. When a training input-output example is presented to the system, the back-propagation algorithm can compute the system output and compares it with the desired output of the training example. The error is propagated backwards through the network from the output layer to the input layer. The neuron activation functions are modified as the error is propagated. To decide the necessary modifications, the back-propagation algorithm differentiates the activation functions of the neurons. Figure 12 depicts an ANFIS topology [14].

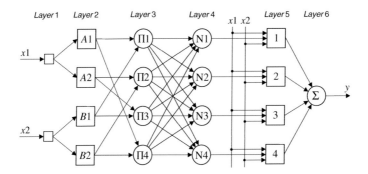

Fig. 12 Depicts an ANFIS topology

5   Design of Fuzzy Logic Controller Tunned by an Expert

The Fuzzy Logic controller proposed in this paper is a Type-2. To produce the membership functions and inference rules for the controller; it is also fundamental to set up the control target: the Quadrotor must be able to keep the requested orientation of its three rotation angles, so that it follows an established reference threshold ($\pm 10°$ around the horizontal for the case of x and y axes and the initial value for the case of z), while it is capable of rejecting external physical disturbances of restrained amplitude applied over its axes. Given that the motors are 90° from each other, it can be assumed that there are two axes with one motor in each side and that the motors that belong to the same axis turn in the same direction. The sensor used is aligned in such a way that a rotation in one physical axis of the Quadrotor provides primarily changes in just one of the measured axes. With this, so as to vary the rotation in one angle (either x or y from Fig. 13) it is just necessary to adjust the output of the motors attached to the perpendicular axis. In the case of adjusting the rotation about z, it is necessary to increase equally the torque of the motors of one axis while reducing in the same value the ones from the other axis. This variation generates the correction in the z axis and avoids variations in x, y or the altitude.

Using the software Solidworks two types of analyzes were performed to compare the behavior of the aluminum (1100 H16) structure against the original plastic (Acrylonitrile butadiene styrene) structure, both of them were run as static analysis but they were useful to understand the different reactions of the structure depending on the material. In the first analysis was simulated a drop test from a height of two meters and an improvement from 6.395 to 1.662 mm was observed in maximum deformation and a reduction of 70 % at maximum effort supported by the structure (See Figs. 14, 15, 16 and 17).

On the other hand, in the second analysis a load of 1 kg at the top was simulated thus covering the possible attachment of a camera or some other accessory to the quadrotor. In this analysis it was observed a lower from 5.916 to 3 mm of maximum

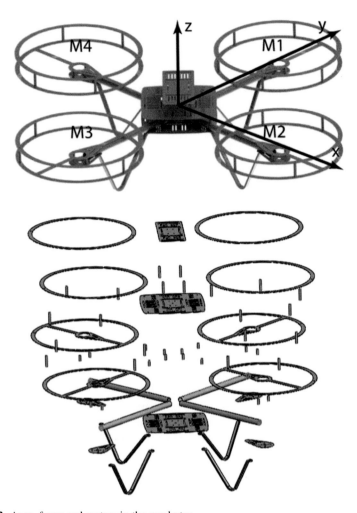

Fig. 13 Axes, frame and motors in the quadrotor

deformation in the metal structure and a reduction of 61 % in the maximum stresses to which the structure is subjected (See Figs. 18, 19, 20 and 21).

From the gyroscope and accelerometer it is possible to acquire the measurements of rotation angle and differentials of this value. The parameters to be considered as inputs and outputs are shown in Table 3.

According with all the experimentation with fuzzy logic controllers, it is determined that 7 membership functions for the rotation of each angle, and 5 for the derivatives will create a sufficient description of the system. This creates the possibility to use a larger number of functions close to the reference value, since a significant part of the control takes part over that interval. The latter implies that while there is a membership function center in zero, the next ones will be close to this

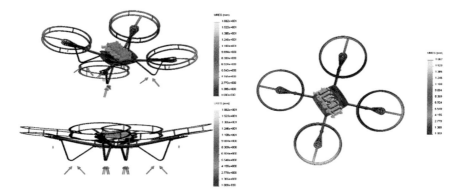

Fig. 14 Drop test aluminum deformation

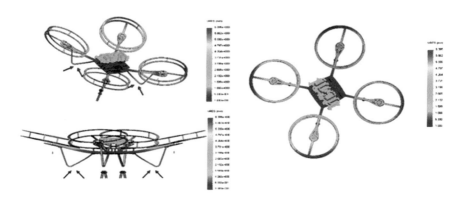

Fig. 15 Drop test plastic deformation

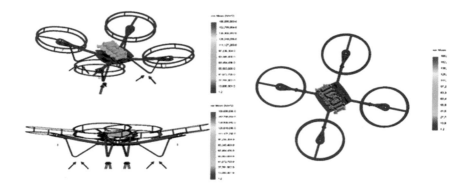

Fig. 16 Drop test aluminum stress

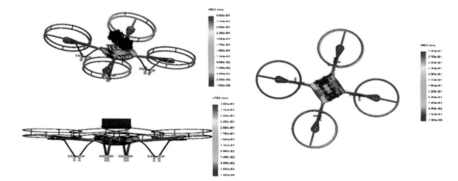

Fig. 17 Drop test plastic stress

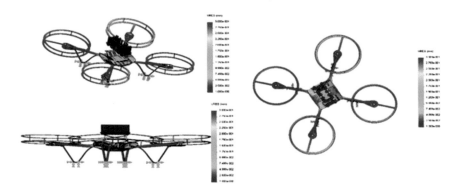

Fig. 18 Load test aluminum deformation

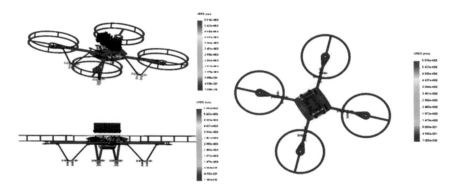

Fig. 19 Load test plastic deformation

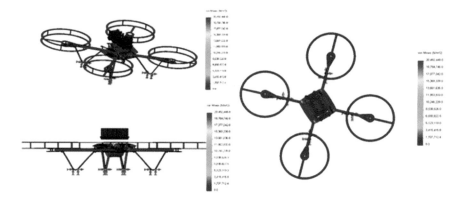

Fig. 20 Load test aluminum stress

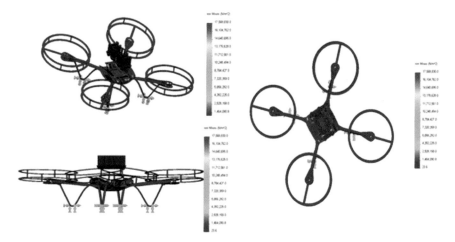

Fig. 21 Load test plastic stress

Table 3 Inputs and outputs of the system

x	Angle in the roll axis
dx	Angular velocity in the roll axis
y	Angle in the pitch axis
dy	Angular velocity in the pitch axis
z	Angle in the yaw axis
dz	Angle in the yaw axis
M1	% of PWM for Motor 1
M2	% of PWM for Motor 2
M3	% of PWM for Motor 3
M4	% of PWM for Motor 4

Fig. 22 Membership
functions for x, y and z

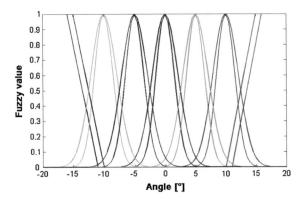

Fig. 23 Membership
Functions for dx, dy and dz

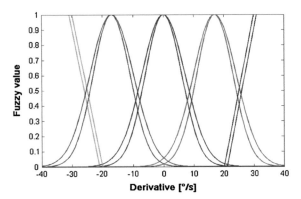

Table 4 Linguistic values of
the MFs

NXL	Negative extra-large
NL	Negative large
NM	Negative medium
NS	Negative small
Z	Zero
PS	Positive small
PM	Positive medium
PL	Positive large
PXL	Positive extra-large

value and the extreme functions will consider an angle value which the Quadrotor
and the controller could not be capable to handle properly anymore generating an
output that will try to correct it. The shape selected for the MFs is the Gaussian. The
MFs are the symmetrical for all the axes because it is expected that they react in the
same manner. The final MFs adjusted are illustrated in Figs. 22 and 23.

The Linguistic variables that describe each membership function (inputs and
outputs) are illustrated in Table 4.

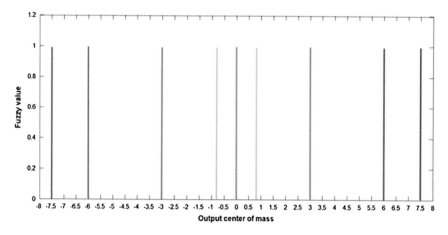

Fig. 24 Membership Functions for M1, M2, M3 and M4

For the output, nine singleton MFs were defined. These MFs represent the PWM percentage value for the motors. The MFs are mounted over a base value of PWM percentage, which is established to be 60. Since the base value was calibrated for each motor, the MFs are the same for the four motors. Figure 24 shows the final MFs of the outputs:

As a result, the physical layout of the Quadrotor explained before (See Fig. 1). Table 4 shows the mapping for the inference rules that relates the input MFs with the output MFs. All the rules addressed the general expression (if-then):
If X is A and Y is B then Z is C

The expert can generate the linguistic rules from the effects generated in the Quadrotor by each motor; two examples are provided for the case of Motor 2 (See Fig. 1).

If the Quadrotor has a negative large angle (Motor 2 is below the horizontal line, Fig. 25) and the derivative is negative medium (it is rotating in the same direction of the negative angle), then Motor 2 needs a Positive Extra-large PWM value in order to brake that rotation and try to compensate the angle (Motor 4 will receive the same value but negative to compensate the torque).

Fig. 25 Large angle and
negative medium derivative

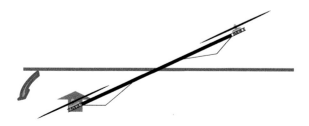

Fig. 26 Inference rule example 2

Table 5 Inference rules

		Derivative				
		NM	NS	Z	PS	PM
Angle	NL	PXL	PXL	PXL	PL	PL
	NM	PXL	PXL	PL	PM	PS
	NS	PXL	PL	PM	PS	NS
	Z	PM	PS	Z	NS	NM
	PS	PS	NS	NM	NL	NXL
	PM	NS	NM	NL	NXL	NXL
	PL	NL	NL	NXL	NXL	NXL

If the Quadrotor has a zero value angle (it is already aligned with the horizontal reference) and its derivative is zero (it is not rotating), then neither Motor 2 nor Motor 4 should receive a new value of PWM (see Fig. 26).

If all the conditions are evaluated, the inference table is obtained. Table 5 illustrates the rules implemented.

The surface generated with the controller designed is shown in Fig. 27. The membership functions proposed allow the system to move smoothly for reaching the position command signal, so the overshoot is reduced.

The motors are affected by the corrective actions of the axis in which they are (x for motors 2 and 4 and y for motors 1 and 3) and by the corrective action in the z axis. Therefore, the final value for each motor will be given by the Eq. (28)

$$
\begin{aligned}
M1 &= BV + Xcorrector + Zcorrector \\
M2 &= BV + Ycorrector - Zcorrector \\
M3 &= BV - Xcorrector + Zcorrector \\
M4 &= BV - Ycorrector - Zcorrector
\end{aligned}
\tag{28}
$$

where BV is the base value of each motor.

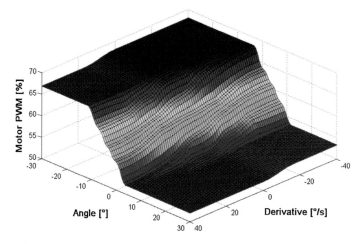

Fig. 27 Surface generated

6 Design of Fuzzy Logic Controller Tunned by an Anfis

If the surface generated by the fuzzy logic controller type 2 (see Fig. 8) is used for training the ANFIS, it is possible to reduce the number of membership functions and linguistic rules. The training method selected is the backpropagation. The ANFIS was trained with 3 membership functions in the angle and 3 membership functions in the derivative. It can be observed that the absolute minimum error reached is equal to 0.4577 which is a good result for flying the Quadrotor (See Fig. 28). If the number of membership functions is increased the error decreased (See Fig. 29) but it is not the goal of this proposal. The target is to decrease the number of membership functions and rules for using lees computational resources. The final domains for the outputs and inputs trained by the ANFIS are presented in Figs. 30, 31 and 32.

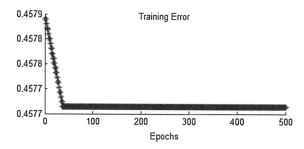

Fig. 28 Absolute minimum error reached with 3 membership functions for the angle and 3 membership functions for the derivative

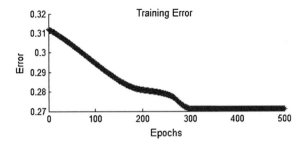

Fig. 29 Absolute minimum error reached with 7 membership functions for the angle and 5 membership functions for the derivative

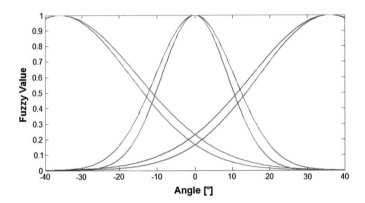

Fig. 30 Angle MFs obtained with ANFIS

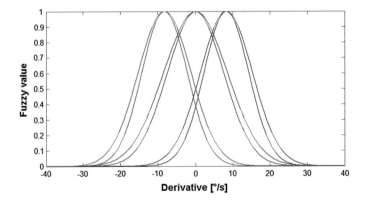

Fig. 31 Derivative MFs obtained with ANFIS

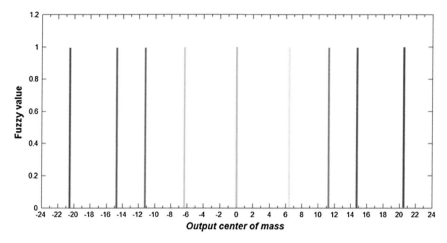

Fig. 32 Output MFs obtained with ANFIS

7 Software Implementation

The controller designed was initially programmed in Matlab in order to obtain the first view of its behavior via the surface generated. The code is shown in Appendix E. Diagram 1 shows the procedure in which the T2 Fuzzy Logic controller is implemented.

The surface obtained from the code is shown in Fig. 25.

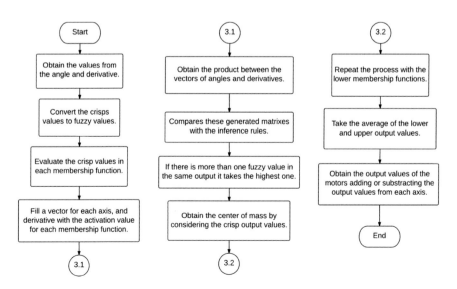

Diagram 1 Implementation of fuzzy logic controller

After confirming the code works as expected, it is translated into the programming language in which the board (BBB) will work. The language selected was Python. The BBB works in a Linux environment, with Debian as operative system. It is connected to the computer via USB. To enter into the CPU of the BBB it is used PuTTY, a Secure Shell (SSH) client. The IP address for the BBB is 192.168.7.2. The program of the controller in Python implemented into the board is shown in Appendix F. The code running tutorial can be found in Appendix C.

The main program used for the project is in appendix G. The following Diagram 2 shows its structure. It starts with a wake up of the sensor. Next, the program takes an average of 500 lectures from the gyroscope in order to calibrate it (the sensor must stay still during this process). Following, the motors are started gradually in order to reduce the peak current. When the PWM of the motors are in the base value, it is established the sample time (needed for obtaining the measures) and a text file is created in which the values of inputs and outputs is written. Right after that, an initial measure is done and the control loop starts.

The measures from the sensor are given in pairs with an I^2C protocol. They are obtained from the following addresses:

Gyroscope :
x: 0x43 and 0x44
y: 0x45 and 0x46
z: 0x47 and 0x48
Accelerometer :
x: 0x3B and 0x3C
y: 0x3D and 0x3E
z: 0x3F and 0x40

After calculating the angles and angular velocities, the program calls the controller to evaluate those measures and generate the output for the motors. The loop continues until the user cancels it. It is important to notice that the adjusting time is crucial for the controller, so the established time for each loop is 22 ms, which was used considering the processing time the controller takes to make a single control loop (Diagram 3).

8 Experimental Results

Experimental results were created with different magnitude of noise added to the position sensor's signal (angle and derivative) and initial position equal to 30° is selected for all the tests. A LabVIEW frontal panel was developed for recording the information for each test; Fig. 33 shows the fontal panel for the Quadrotor.

When the noise was not presented, the error signal from the ANFIS tuned controller is better than the controller tuned by the expert. Figure 34 depicts the response for the controller tuned by the ANFIS (a) and tuned by the expert (b).

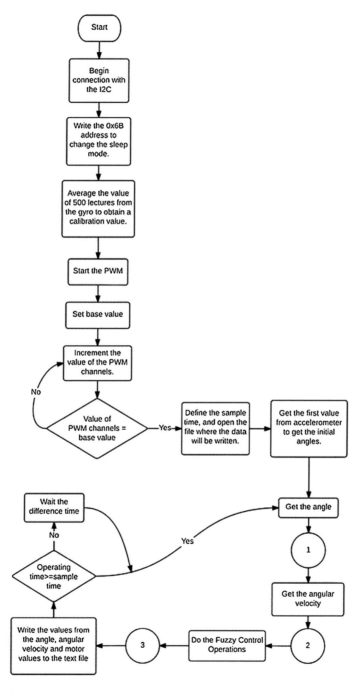

Diagram 2 Main program

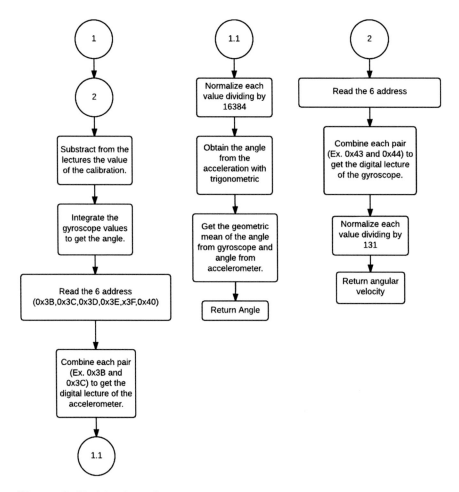

Diagram 3 Obtaining the angle

It is observed that the fuzzy logic controller tuned by ANFIS the error is inside a band (±10°), but the system tuned by the expert some ripples occurred at 22 s.

When noise is added in both angle (±5°) and derivative (±10/s degrees), both controllers can deal with this noise level. Figure 35 shows how the response in the controller tuned by the ANFIS is degraded and the expert response can tolerate the noise in the band (±10°).

When noise is added in both angle (±25°) and derivative (± 50/s degrees), the fuzzy logic controller tuned with ANFIS gives a better response than the controller tuned by the expert (See Fig. 36) which has big ripples in the response.

It can be observed that the fuzzy logic controller tuned by ANFIS gives good response with no-noise and high level of noise but the response with medium level of noise is improved by the controller tuned by the expert because the number of

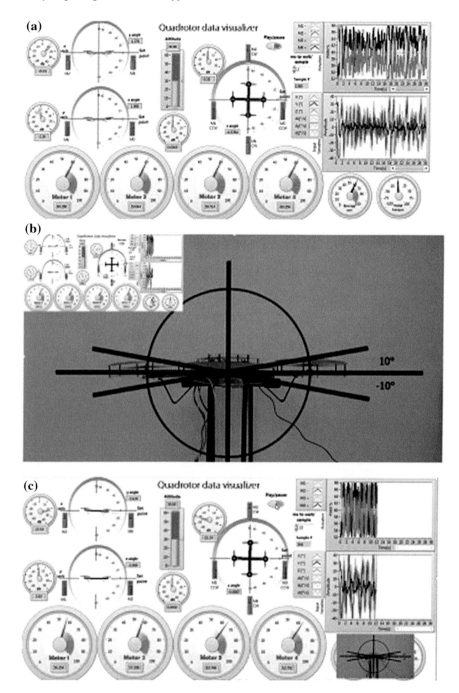

Fig. 33 LabVIEW frontal panel for the quadrotor (**a**), quadrotor running (**b**) and frontal panel and quadrotor (**c**)

Fig. 34 No noise included in
the system, controller tuned
by ANFIS (**a**) and tuned by
the expert (**b**)

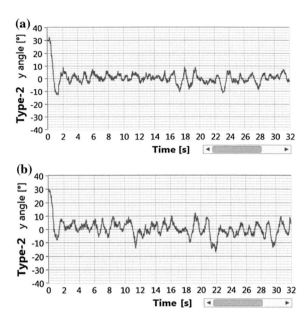

Fig. 35 Noise included in the
system, controller tuned by
ANFIS (**a**) and tuned by the
expert (**b**) medium level of
noise

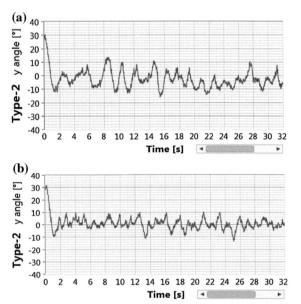

Fig. 36 No noise included in the system, controller tuned by ANFIS (**a**) and tuned by the expert (**b**) high level of noise

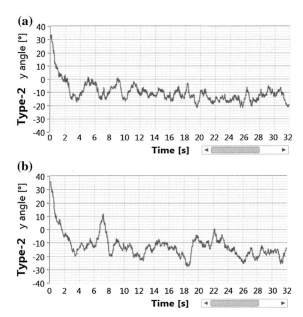

Table 6 Processing time for each fuzzy logic controller in (ms)

	Type 2 7&5	ANFIS T2 3&3
	21.1	14.4
	21.8	14.7
	21.7	14.6
	20.9	14.7
	20.3	15.2
	21.4	13.6
	21.0	13.6
	20.0	15.4
	21.5	14.4
	21.6	14.7
Average	21.1	14.5

membership functions generate smooth transitions. When more membership functions are implemented, the transition with medium level of noise is smoother than fuzzy logic controller tuned by ANFIS.

Table 6 shows how the processing time changes when a different fuzzy logic controller is implemented in the digital micro-controller. The BeagleBone Black (BBB) was the microprocessor selected. A list of some specifications of the board is presented below:

- Processor: AM3359 ARM Cortex-A8
- Speed Processor: 1 GHz
- Memory: 512 MB DDR3 (800 MHz x 16), 2 GB on-board storage using eMMC
- Digital pins: 65
- Analog pins: 7
- PWM pins: 8

In this case, it is clear that Type-2 optimized by ANFIS needs a lower processing time than Type-2 tuned by the expert. It is important to mention that fuzzy logic controller type 2 tuned by an expert is composed of 7 membership functions for the angle and 5 membership functions for the derivative, while the ANFIS tuned controller is composed only by 3 membership functions for the angle and 3 membership functions for the derivative.

9 Conclusions

- Tuning a controller by trial and error is not as efficient as implementing a software tuning tool, given the large amount of tests that were made for this. After optimizing with the ANFIS toolkit, the performance of the generated controller was similar, but with some variations caused by the existence of error in the output surface that was used in order to tune it. However, it was clear that the processing time diminished even more than using the Type-1 controller, providing a powerful advantage of implementing a controller tuned with this tool.
- If the process was made the other way round, that is, adjusting the desired output surface instead of the input membership functions, the controller could increase its performance given that is the output what relates directly to the behavior. In such a case, ANFIS could be an excellent tool to obtain the required input Membership Functions.
- The analysis done in Solidworks for comparison of materials in the structure showed a substantial improvement with aluminum at reducing deformation and maximum efforts supported by the structure for more than 50 % in both analyzes, confirming that aluminum was a better material for building the quadrotor.
- The development of a LabVIEW™ frontal panel allowed a complete and exhaustive analysis of the results, and along with an attractive visual interface with the use of images, needle indicators and graph charts to display the results, it was possible to tune the controller until obtaining the desired results.

References

1. Altug, E.S.: Adaptive Control of a Tilt—Roll Rotor Quadrotor UAV. IEEE, Orlando, FL (2014)
2. Bhatkhande, P.H.: Real Time Fuzzy Controller for Quadrotor Stability Control. IEEE, Beijing (2014)
3. Bouabdallah, S.: Design and Control of Quadrotors With Application to Autonomous Flying. Ecole Polytechnique Federale de Lausanne, Lausanne (2007)
4. Brito, J.M.: Quadrotor Prototype. Dissertação para obtenção do Grau de Mestre. Universidad Técnica de Lisboa, Lisboa (2009)
5. Castillo, P.G.: Robustness with respect to delay uncertainties of a predictor-observer based discrete-time controller. IEEE, San Diego, CA (2006)
6. Coza, C.M.: A New Robust Adaptive-Fuzzy Control Method Applied to Quadrotor Helicopter Stabilization. Fuzzy Information Processing Society, 2006. NAFIPS 2006. Annual meeting of the North American, pp. 454–458 (2006)
7. Coza, C.M.: A New Robust Adaptive-Fuzzy Control Method Applied to Quadrotor Helicopter Stabilization, p. 475. IEEE (2006)
8. Drossos, C.: How do I choose membership functions in a fuzzy system? Recuperado el 16 de 11 de 2014 (18 de 12 de 2013). de ResearchGate: http://www.researchgate.net/post/How_do_I_choose_membership_functions_in_a_fuzzy_system
9. Flynn, E.P.: Low-Cost Approaches to UAV Desgin Using Advanced Manufacturing Techniques. IEEE (2013)
10. Guenard, N.C.: Visual Navigation of a Quadrotor Aerial Vehicle. IEEE, St. Louis (2009)
11. Hoffman, G.R.: The Standford Testbed of Autonomous Rotorcraft for Multiagent Control (STARMAC). Standford University, California (2004)
12. Hoffmann, G.H.: Quadrotor Helicopter Flight Dynamics and Control: Theory and Experiment. AIAA Guidance, Navigation and Control Conference and Exhibit. American Institute of Aeronautics and Astronautics, South Carolina (2007)
13. Ilhan, I., Karakose, M.: Type-2 fuzzy based quadrotor control approach. In Control Conference (ASCC), 2013 9th Asian (pp. 1–6). IEEE (2013)
14. Jang, J.S.R.: ANFIS: adaptive-network-based fuzzy inference systems. IEEE Trans. Syst Man Cybern. 23, 665–685 (1993)
15. Karnik, N.M.: Type-2 fuzzy logic systems. IEEE Trans. Syst Man Cybern. 643–658 (1999)
16. Puls, T., Kemper, M., Küke, R., Hein, A.: GPS-based Position Control and Waypoint Navigation System for Quadrocopters. In Intelligent Robots and Systems, IROS 2009. IEEE/RSJ InternationalConference. 3374–3379 (2009)
17. Kirli, A.R.: Self tuning fuzzy PD application on TI TMS320F28335 for an experimental stationary quadrotor. Yildiz Tech. Univ. (2010)
18. Leishman, L.: Principles of Helicopter Aerodynamics (Cambridge Aerospace Series). Cambridge, MA (2000)
19. Mahony, R.: Multirotor aerial vehicles: modeling, estimation, and control of quadrotor. IEEE Robot. Autom. Mag. (2012)
20. Maroto, R.: Ecuaciones para la sintonización de controladores PID con acción derivativa aplicada a la señal retroalimentada. Ing Téc, Universidad de Costa Rica, Facultad de Ingeniería, Escuela de Ingeniería Eléctrica. (2007)
21. Mendel, J.R.: Interval type-2 fuzzy logic systems made simple. IEEE Trans. Fuzzy Syst. 808–821 (2006)
22. Mendel, J.: Type-2 fuzzy sets and systems: an overview. IEEE Comput. Intel. Mag. 20–29 (2007)
23. Nguyen, H.K.: Fuzzy logic, logic programming, and linear logic. IEEE 546–550 (1996)
24. Pounds, P.E.: Design, Construction and Control of a Large Quadrotor Micro Air Vehicle (Doctoraldissertation, Australian National University) (2007)

25. Gao, Q., Yue, F., Hu, D.: Research of stability augmentation hybrid controller for quadrotor UAV. In: Control and Decision Conference (2014 CCDC), The 26th Chinese pp. 5224–5229. IEEE (2014)
26. Rollins, L.: Robust Control Theory. Carnegie Mellon University, Pittsburgh (1999). http://users.ece.cmu.edu/~koopman/des_s99/control_theory/
27. Roy, A.M.: Fuzzy logic, neural networks, and brain-like learning. In Neural Networks, International Conference, vol. 1, pp. 522–527. IEEE (1997)
28. Santos, M.L.: Intelligent fuzzy controller of a quadrotor. IEEE 141–146 (2010)
29. Sheikhpour, S., Shouraki, S.B.: A Model-Based Fuzzy Controller Using the Parallel Distributed Compensation Method for Quadrotor Attitude Stabilization. InElectrical Engineering (ICEE), 2013 21st IranianConference, pp. 1–6. IEEE (2013)
30. Tournier, G.: Six Degree of Freedom Estimation Using Monocular Vision and Moire Patterns. MIT, Cambridge (2006)
31. Wu, H.-J., Su, Y.-L., Lee, S.J.: A fast method for computing the centroid of a type-2 fuzzy set. IEEE Trans. Syst. Man Cybern. **42**, 764–777 (2012)
32. Yiming, Y.F.: Rule Based Fuzzy Logic Inferencing. IEEE (1994)

Mobile Robot with Movement Detection Controlled by a Real-Time Optical Flow Hermite Transform

Ernesto Moya-Albor, Jorge Brieva and Hiram Eredín Ponce Espinosa

Abstract This chapter presents a new algorithm inspired in the human visual system to compute optical flow in real-time based on the Hermite Transform. This algorithm is applied in a vision-based control system for a mobile robot. Its performance is compared for different texture scenarios with the classical Horn and Schunck algorithm. The design of the nature-inspired controller is based on the agent-environment model and agent's architecture. Moreover, a case study of a robotic system with the proposed real-time Hermite optical flow method was implemented for braking and steering when mobile obstacles are close to the robot. Experimental results showed the controller to be fast enough for real-time applications, be robust to different background textures and colors, and its performance does not depend on inner parameters of the robotic system.

Keywords Optical flow · Hermite transform · Movement detection · Navigation system · Differential method

1 Introduction

Movement detection and characterization in a scene is a relevant task in vision systems used in robotic applications controlled by visual features. The optical flow approach allows to obtain the displacement (magnitude and direction) of each pixel in the image in a given time. The quality and precision of these displacements in real time will determine the performance of the control system. The optical flow

E. Moya-Albor · J. Brieva (✉) · H.E. Ponce Espinosa
Faculty of Engineering, Universidad Panamericana, 03920 Mexico City, Mexico
e-mail: jbrieva@up.edu.mx

E. Moya-Albor
e-mail: emoya@up.edu.mx

H.E. Ponce Espinosa
e-mail: hponce@up.edu.mx

© Springer International Publishing Switzerland 2016
H.E. Ponce Espinosa (ed.), *Nature-Inspired Computing for Control Systems*,
Studies in Systems, Decision and Control 40, DOI 10.1007/978-3-319-26230-7_9

231

problem has been treated widely in the literature. Some of them are easily implemented and extended to real time applications but with low accuracy, as [12], and others have fine implementations and high accuracy but most are impossible to use in real time applications by hardware limitations and high processing time, as [25] and [21]. In this chapter, we propose an algorithm in real time, using the optical flow constraint equation of Horn and Schunck [12], as the differential approach most used in real time applications, with improvements in accuracy due to the proposal using the Hermite transform. The Hermite transform is a biologically inspired image model, allowing fully describe the significant visual features in digital images, and together with the Horn and Schunck differential approach provide a real-time optical flow method more robust to noise and to intensity changes in the image sequence.

Computer vision can help to sense control variables in a navigation system. This kind of sensors is not conventional and is less used than others in navigation systems. Particularly, optical flow models can help to measure magnitude and direction displacement in a particular scene. In literature, there are several applications using optical flow especially in navigation problems: for aeronautical applications was proposed an aircraft maneuvering controlled by translational optical flow in [27], in [14] they used optical flow based on block-matching algorithm, image-based lane departure warning (LDW) system using the Lucas-Kanade optical flow and the Hough transform methods was proposed in [28], application to a lunar landing scenario using optical flow sensing was addressed in [30], for Power Assisted Wheelchair in [19] and Optical Flow Based Plane Detection for Mobile Robot Navigation in [32], and in [37] they used a controller design based on real time optical flow and image processing applied in a quadrotor and a Visual Tracking of a Moving Target by a camera mounted on a robot. Other applications not directly related to navigation systems are presented in literature as Counting Traffic [1], real time velocity estimation based on OF and disparity [11], and optical-flow-based altitude sensor and its integration with flapping-wing flying microrobot [7].

This chapter is organized as follows: Sect. 2 gives the basic concepts about the biological inspired model. The principles of optical flow methods are described in Sects. 3 and 4 presents our approach to compute optical flow in real time and some tests to validate our approach. Section 5 describes the implementation of our approach in the navigation of mobile robots. A conclusion is finally reported in Sect. 6.

2 Hermite Transform: A Biologically Inspired Model of Digital Images

This section introduces the Hermite Transform as a bio-inspired model for represent relevant perceptive features in digital images. First, we list the representative bio-inspired image models. Then, the basic concepts of the Hermite Transform and its relationship with the human vision system is presented. Finally, we show a steered version of the Hermite Transform, that allows describe oriented patterns.

2.1 Bio-inspired Image Models

A large number of image processing algorithms has been inspired on the human visual perception systems. The principal task that these approaches must resolve is to realize a good description of the image to identify, reconstructing and tracking the objects in order to interpret the scene. Different methodologies can be divided into two general categories: those requiring strong prior knowledge and those requiring weak or no prior knowledge. In the second group the most important techniques are based on psychovisual human studies [29] and biological models [34]. The first one proposed by Gestalt Psychologists point out the importance of visual organization or perceptual grouping in the description of the scene, particularly the objects in the scene cannot be processed separately. They define concepts as proximity, continuity, homogeneity as principles to implement perception rules for processing the image. The main difficulty of these approaches is the implementation of the perception rules that must be particular to each application.

Approaches based on biological models seem more adequate for image processing due to their ability to generalize to all kind of scenes. They are inspired directly on the response of the retina and the ganglion cells to light stimuli (see Fig. 1) A first model proposed by Ernst Mach recognize that not just light intensities but intensity changes influence what we see (i.e., derivatives and sum of second derivatives in the space). The Gabor model, proposed by [26], represents the receptive fields of the visual cortex through Gaussian modulated with complex exponentials. Like the receptive fields, the Gabor functions are spatially local and consist of alternating bands of excitation and inhibition in a decaying envelope.

In 1987, Young [36] proposed a receptive field model based on the Gaussian and its derivatives. These functions, like the Gabor, are spatially local and consist of alternating regions of excitation and inhibition within a decaying envelope. Young

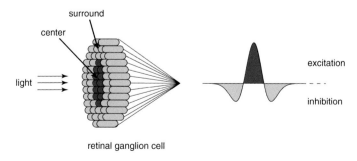

Fig. 1 Diagram of ganglion cells showing the biological inspiration for image processing

showed that Gaussian derivative functions more accurately model the measured receptive field data than the Gabor functions. This functions can be interpreted as the product of Hermite polynomials and a Gaussian window. It allows to decompose the image based on Hermite polynomials by defining a biological model of the measured receptive field data in the Human Vision System (HVS) [34–36].

The Hermite Transform uses a Gaussian window to localize and analyze the visual information where from a perceptual standpoint, the Gaussian window is a good model of the overlapping receptive fields found in physiological experiments [26]. Furthermore, the associate polynomials involves Gaussian derivative operators found in psychophysical models of the HVS [4, 17]. Finally, the operators used in the Hermite Transform provide a natural agreement with the theory scale-space [13] where the Gaussian window minimizes the uncertainty of the product of the spatial and frequency domain [33].

2.2 The Hermite Transform

The Hermite Transform is a special case of polynomial transform and was introduced by Martens in 1990 [18]. First, the original image $L(x, y)$ with coordinates (x, y) is located at various positions multiplying it by a window function $v^2(x - x_0, y - y_0)$ at positions (x_0, y_0) that conform a sampling lattice S. Then, the local information for each analysis window is expanded in terms of a family of orthogonal polynomials $G_{m,n-m}(x, y)$ where m and $(n - m)$ denote the analysis order in x and y direction respectively.

The Hermite Transform of an image can be computed by a convolution of the image $L(x, y)$ with the filter functions $D_{m,n-m}(x, y)$ obtaining the cartesian Hermite coefficients $L_{m,n-m}(x, y)$:

$$L_{m,n-m}(x_0, y_0) = \int\limits_{-\infty}^{\infty} \int\limits_{-\infty}^{\infty} L(x, y) D_{m,n-m}(x_0 - x, y_0 - y) dx dy \tag{1}$$
$$n = 0, 1, \ldots, \infty \quad m = 0, 1, \cdots, n$$

The filter functions $D_{m,n-m}(x, y)$ are defined by polynomials $G_{m,n-m}(x, y)$ that are orthogonal with respect to an analysis window $v^2(x, y)$

$$D_{m,n-m}(x, y) = G_{m,n-m}(-x, -y) v^2(-x, -y), \tag{2}$$

where $v(x, y) = \frac{1}{\sigma\sqrt{\pi}} \exp\left(-\frac{(x^2 + y^2)}{2\sigma^2}\right)$ is a Gaussian window with normalization factor for a unitary energy for $v^2(x, y)$.

The polynomials orthonormal with respect to $v(x,y)^2$ can then be written as

$$G_{m,n-m}(x,y) = \frac{1}{\sqrt{2^n m!(n-m)!}} H_m\left(\frac{x}{\sigma}\right) H_{n-m}\left(\frac{y}{\sigma}\right), \tag{3}$$

where $H_n(x)$ are the Hermite polynomials given by Rodrigues' formula:

$$H_n(x) = (-1)^n \exp^{(x^2)} \frac{d^n}{dx^n} \exp^{(-x^2)} \quad n = 0,1,2,\ldots \tag{4}$$

and $H_n\left(\frac{x}{\sigma}\right)$ represents the generalized Hermite polynomials with respect to the Gaussian function (with variance σ^2).

In Fig. 2, we show the Hermite polynomials $H_n(x)$ for $n = 0,1,2,3,4,5$.

The corresponding analysis Hermite filters are separable both spatial and polar and they can be expressed by

$$D_{m,n-m}(x,y) = D_m(x)D_{n-m}(y), \tag{5}$$

where the one-dimensional Hermite filters can be computed by

$$D_n(x) = \frac{(-1)^n}{\sqrt{2^n n!}} \frac{1}{\sigma\sqrt{\pi}} H_n\left(\frac{x}{\sigma}\right) \exp^{-\frac{x^2}{\sigma^2}}. \tag{6}$$

Fig. 2 The Hermite polynomials $H_n(x)$ for $n = 0,1,2,3,4,5$

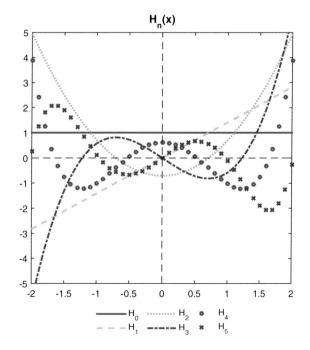

Figure 3a shows the Hermite filters $D_{m,n-m}(x,y)$ for $N = 3$ ($n = 0, 1, \ldots, N$ and $m = 0, 1, \ldots, n$).

In Fig. 4a, we obtained the cartesian Hermite coefficients of the *House* test image [10] for $N = 3$ ($n = 0, 1, 2, 3$).

(a)   **(b)**

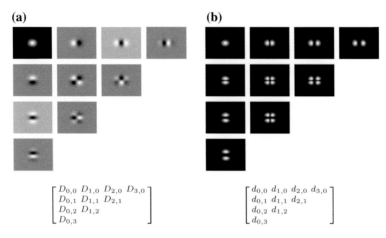

$$\begin{bmatrix} D_{0,0} & D_{1,0} & D_{2,0} & D_{3,0} \\ D_{0,1} & D_{1,1} & D_{2,1} \\ D_{0,2} & D_{1,2} \\ D_{0,3} \end{bmatrix} \qquad \begin{bmatrix} d_{0,0} & d_{1,0} & d_{2,0} & d_{3,0} \\ d_{0,1} & d_{1,1} & d_{2,1} \\ d_{0,2} & d_{1,2} \\ d_{0,3} \end{bmatrix}$$

Fig. 3 **a** An ensemble of the Hermite filters $D_{m,n-m}(x,y)$ and **b** their Fourier transform spectrum $d_{m,n-m}(\omega_x, \omega_y)$ (*right*) for $N = 3$ ($n = 0, 1, \ldots, N$ and $m = 0, 1, \ldots, n$) [22]

(a)   **(b)**

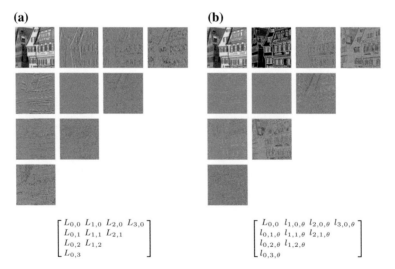

$$\begin{bmatrix} L_{0,0} & L_{1,0} & L_{2,0} & L_{3,0} \\ L_{0,1} & L_{1,1} & L_{2,1} \\ L_{0,2} & L_{1,2} \\ L_{0,3} \end{bmatrix} \qquad \begin{bmatrix} L_{0,0} & l_{1,0,\theta} & l_{2,0,\theta} & l_{3,0,\theta} \\ l_{0,1,\theta} & l_{1,1,\theta} & l_{2,1,\theta} \\ l_{0,2,\theta} & l_{1,2,\theta} \\ l_{0,3,\theta} \end{bmatrix}$$

Fig. 4 **a** An ensemble of the cartesian Hermite coefficients and **b** the Steered Hermite coefficients of the *House* test image for $N = 3$ ($n = 0, 1, 2, 3$) [20]

2.3   The Steered Hermite Transform

The Steered Hermite Transform (*ST*) is a version of the cartesian Hermite coefficients that is obtained by rotating of the cartesian coefficients towards an estimated local orientation, according to a criterion of maximum oriented energy at each window position.

The Hermite filters $D_{m,n-m}(x, y) = D_m(x)D_{n-m}(y)$ are separable in space, and its Fourier transform can be expressed in polar coordinates. If $\omega_x = \omega \cos(\theta)$ and $\omega_y = \omega \sin(\theta)$, then

$$d_m(\omega_x)d_{n-m}(\omega_y) = g_{m,n-m}(\theta) \cdot d_n(\omega), \tag{7}$$

where $d_n(\omega)$ is the Fourier transform of each filter function, which expresses radial frequency selectivity of the nth derivative of the Gaussian but with a radial coordinate r for x, and $g_{m,n-m}(\theta)$ expresses the directional selectivity of the filter

$$g_{m,n-m}(\theta) = \sqrt{\binom{n}{m}} \cos^m \theta \cdot \sin^{n-m} \theta \tag{8}$$

The orientation feature of the Hermite filters explains why they are products of polynomials with a radially symmetric window function (Gaussian function). The $N+1$ Hermite filters of order n form a steerable basis for each individual filter of order n. Filters of increasing order n analyze successively higher radial frequencies (Fig. 3b), and filters of the same order n and different (directional) index m distinguish between different orientations in the image. Note that the radial frequency selectivity $d_n(\omega)$ is the same for all $N+1$ filters of order n and that these filters differ only in their orientation selectivity [18]. The resulting filters can be interpreted as directional derivatives of a Gaussian function.

For local 1D patterns, the Steered Hermite Transform provides a very efficient representation. This representation consists of a parameter θ that indicates the orientation of the pattern, and a small number of coefficients that represent the profile of the pattern perpendicular to its orientation. For a 1D pattern with orientation θ, the following relation holds:

$$l_{m,n-m,\theta}(x, y) = \begin{cases} \sum_{k=0}^{n} (L_{k,n-k}(x, y))(g_{k,n-k}(\theta)), & m = 0 \\ 0, & m > 0, \end{cases} \tag{9}$$

where $l_{m,n-m,\theta}(x, y)$ are Steered Hermite coefficient for the local angle θ.

This means that a Steered Hermite Transform offers a way to describe 1D patterns explicitly on the basis of their orientation and profile [31].

In Fig. 4b, we steer the cartesian Hermite coefficients of the corresponding coefficients of Fig. 4a according to maximum energy direction. It is noticeable that the energy is concentrated in only three coefficients (first row of Fig. 4b), which represent the orientation of the different structures of the image.

3 Optical Flow Problem

This section describes the optical flow problem focused in the differential approaches used to obtain the displacements of the objects in a sequence of image. We present the seminal differential methods of Horn and Schunck, and Lucas and Kanade to highlight the advantages of global and local proposal. Later, we mention the additional constraints of recent approaches to improve robustness of classical methods.

3.1 Definition

The optical flow (OF) is defined as a two-dimensional distribution of apparent velocities that can be associated with variations of brightness patterns in a sequence of images [9]. The obtained vector field shows the displacement of each pixel at some time in the image sequence and represents the motion of objects in the scene or the relative motion of the sensor.

To illustrate the optical flow concept, we show in Fig. 5 the vector field obtained from the frames 43 and 44 of the *Cameramotion* sequence [15], where the apparent

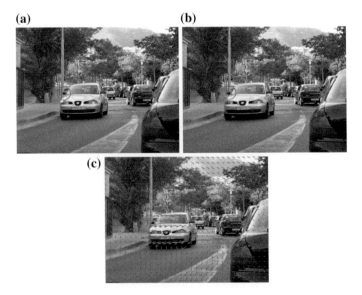

Fig. 5 Optical flow representation in a sequence of images [20]. **a** Frame 43 of the *Cameramotion* sequence. **b** Frame 44 of the *Cameramotion* sequence. **c** Optical flow result

motion of the car of the left is highlighted by the side view mirror within the red circle.

The vector of displacements obtained in two successive images can be observed by:

- The motion of objects in the scene.
- The relative movement between the observer and the scene.
- Variations in scene illumination.

Various methods have been proposed to calculate the optical flow in image sequences since 1980. Barron et al. [2] performed the first classification of the existing optical flow methods into four groups:

- Differential methods
- Region-based matching methods
- Energy-based methods
- Phased-based methods

The benchmarks of Barron et al. [2] in 1994 and Galvin et al. [8] in 1998 showed that the differential methods and phase-based methods were the techniques with better performance. Their study emphasizes the measurement accuracy and concludes that the most accurate methods are the differential proposals, where methods using global smooth constraint appear to produce visually attractive flow fields.

Differential techniques compute image velocity from spatiotemporal derivatives of image intensities or filtered versions of the image. The first approaches use first order derivatives and are based on image translation assuming intensity conservation. Second order differential methods use the Hessian of the intensity function to constrain 2D velocity. There are global and local first- and second-order methods, where global methods use the additional global constraint to compute dense optical flows over large image regions. Local methods use normal velocity information in local neighborhoods to perform a least squares minimization to find the best fit for the vertical and horizontal components of displacement [3].

3.2 Differential Optical Flow Methods

The differential methods are based on the work by Horn and Schunck (HS) [12], which proposes the *Constant Intensity Constraint* assuming that the intensities of the pixels of the objects remain constant:

$$L(X + W) - L(X) = 0, \tag{10}$$

where $L(X, t)$ is an image sequence, with $X = (x, y, t)^{\mathrm{T}}$ representing the pixel location within a rectangular image domain Ω; $W := (u, v, 1)^{\mathrm{T}}$ is a vector that defined the displacement u and v of each a pixel at position (x, y) within the sequence of images at a time t to a time $(t + 1)$ in the directions x and y; respectively.

Considering linear displacements (10) is expanded by Taylor's series, obtaining the *Optical Flow Constraint* equation

$$W^{\mathrm{T}}(\nabla_3 L) = 0, \tag{11}$$

where $\nabla_3 L := (L_x, L_y, L_t)^{\mathrm{T}}$ and $L_* := \frac{\partial L}{\partial *}$.

The optical flow constraint Eq. (11) incorporates the *Ill-posed Aperture Problem*, and it is not sufficient to determine the two unknown functions u ans v uniquely, i.e., in homogeneous areas where movement is observed locally, only the normal component of the movement can be estimated. To overcome this problem, some additional constraints are required.

In 1981, Horn and Schunck [12] proposed the *Smoothness Constraint* that assumed that the apparent speed of the intensity pattern in the image varies smoothly; that is, neighboring points of the objects have similar velocities. To recover the optical flow the approach of Horn and Schunck minimizes a energy functional of the type

$$E_{HS}(W) = \int_\Omega (W^{\mathrm{T}}(\nabla_3 L \nabla_3 L^{\mathrm{T}})W + \alpha |\nabla W|^2)dX. \tag{12}$$

The functional of (12) is a global differential method which allows to obtain dense vector fields. The first term is a data term requiring that the optical flow constraint equation is fulfilled, while the second term penalizes deviations from smoothness. The smoothness term $|\nabla W|^2 = |\nabla u|^2 + |\nabla v|^2$ is called *regularizer* and the positive smoothness weight α is the regularization parameter. One would expect that the specific choice of the regularizer has a strong influence on the result.

One classic local differential approach that minimizes (11) was proposed by Lucas and Kanade (LK) [16] in 1981. They consider the flow constant within a neighborhood ρ (Gaussian function K_ρ of standard deviation ρ), and determine the two constants u and v at a point (X, t) using a weighted least squares approach: a weighted least squares approach:

$$E_{LK}(W) = W^{\mathrm{T}} J_\rho(\nabla_3 L)W, \tag{13}$$

where

$$J_\rho(\nabla_3 L) := K_\rho \circledast (\nabla_3 L \nabla_3 L^{\mathrm{T}}) \tag{14}$$

and $\circledast$ represents the convolution operator.

The local differential methods have the advantage of being robust against noise, but with the disadvantage that they do not get dense flows and an additional interpolation step is required to recover the displacement vector in each pixel of the scene.

To improve robustness to global differential methods, recent approaches have emerged and have suggested some additional constraints [6].

An immediate problem of (12) is that the intensity does not always remain constant from one image to another, therefore an independent intensity change measure is required. In [23, 24] a *Constant Gradient Constraint* (generally used in local methods to handle the aperture problem) is proposed:

$$\nabla L(X + W) - \nabla L(X) = 0. \tag{15}$$

Different approaches have been proposed to obtain accurate fields of displacements, one such methods is a combination of global (Horn and Schunck) and local (Lucas and Kanade) methods to generate dense flow fields, and robust against noise [5, 6].

Another proposal is a spacetime formulation performing a convolution with a three-dimensional Gaussian function and considering soft flows in the temporal direction. This improves robustness to noise and in those cases where the linearization of (10), and in consequence (11), is not valid for large displacements, multi-resolution strategies are used [5]. One option is to propose functionals that combine local constraints (intensity constraint and gradient constraint), a spatiotemporal smoothness constraint and a multi-scale approach [25].

In this sense, the Hermite Transform can be used to define a functional that performs a polynomial decomposition using the Steered Hermite Transform of two consecutive images in a short period of time. This decomposition represents the local characteristics of images from an perceptual approach within a multi-resolution scheme [21].

4 Real Time Hermite Optical Flow Method

This section defines the real-time optical flow method using the Hermite transform, the local constraints of Horn and Schunck approach are defined using the zero order and steered Hermite coefficients as local descriptors of visual features of the images.

4.1 Optical Flow Using the Hermite Transform

The main disadvantages of the recent differential methods for real time applications are their high computational time and difficult to implement. On the other hand, the Horn and Shunk [12] approach is a fast method with low implementation complexity. The main disadvantage of the Horn and Shunck method is its low accuracy. In those applications that require an approximation of the displacements, such method is sufficient in most cases.

We proposed a modified version of Horn and Shunck approach that allows to increase the accuracy of the optical flow by using the Hermite Transform as a biological image model. An expansion of the *Constant Intensity Constraint* (10), with the incorporation of the *Steered Hermite Coefficient Constraint* of the Hermite Transform, is defined as follows

$$\left[L_0(X+W) - L_0(X)\right] + \gamma\left[\sum_{n=1}^{N} l_{n,\theta}(X+W) - \sum_{n=1}^{N} l_{n,\theta}(X)\right] = 0, \qquad (16)$$

where γ is a weight parameter.

Equation (16) includes the local characteristics of the both images that represent the homogenous regions (low frequencies) in the zero order coefficients (L_0) and the edges, textures and complex structures (high frequencies) in the steered Hermite coefficients ($l_{n,\theta}$). In Fig. 6 we show the L_0 and $l_{n,\theta}$ Hermite coefficients from the image 42 and 43 of *Cameramotion* sequence, where there is a displacement (W) of the pixels in the position (X) between an image at time t and another image at time $(t+1)$.

Considering linear displacements of (16) and expanded by Taylor series, we obtain the *Optical Flow Hermite Constraint* equation

$$\begin{aligned}
&\left[u\frac{\partial L_0(X)}{\partial x} + v\frac{\partial L_0(X)}{\partial y} + \frac{\partial L_0(X)}{\partial t}\right] \\
&+ \gamma\sum_{n=1}^{N}\left[u\frac{\partial l_{n,\theta}(X)}{\partial x} + v\frac{\partial l_{n,\theta}(X)}{\partial y} + \frac{\partial l_{n,\theta}(X)}{\partial t}\right] = 0.
\end{aligned} \qquad (17)$$

The one-dimensional spatial derivatives can be reduced considering that the one-dimensional Hermite coefficients are achieved by the inner product between the signal located by the Gaussian window and the Hermite polynomials:

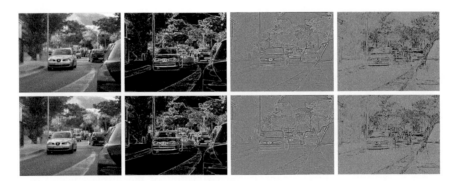

Fig. 6 Hermite coefficients from images 42 (*first row*) and 43 (*second row*) of *Cameramotion* sequence [20]. In first row are shown the coefficients $L_{0,0}(X)$, $l_{1,\theta}(X)$, $l_{2,\theta}(X)$ and $l_{3,\theta}(X)$ at time t. Second row shows the coefficients $L_{0,0}(X+W)$, $l_{1,\theta}(X+W)$, $l_{2,\theta}(X+W)$ and $l_{3,\theta}(X+W)$ at time $t+1$

$$L_k = \left\langle L(x), H_k\left(\frac{x}{\sigma}\right)\right\rangle. \tag{18}$$

It can be demonstrated that the derivative of the Hermite coefficients holds [21]:

$$L_k = L^{(k)}(x) = \frac{\partial^k L(x)}{\partial^k}, \tag{19}$$

and therefore,

$$\frac{\partial L_0(X+W)}{\partial x} = L_{1,0}(X+W),$$

$$\frac{\partial}{\partial x}\sum_{n=1}^{N} l_{n,\theta}(X) = \sum_{n=1}^{N} l_{n,\theta_{(m)+1}}(X), \tag{20}$$

and in a similar way for the derivatives of y.

The *Optical Flow Hermite Constraint* equation can be rewritten as:

$$\left[uL_{0,1}(X) + vL_{1,0}(X) + \frac{\partial L_0(X)}{\partial t}\right]$$

$$+\gamma\sum_{n=1}^{N}\left[ul_{n,\theta_{(m)+1}}(X) + vl_{n,\theta_{(n)+1}}(X) + \frac{\partial l_{n,\theta}(X)}{\partial t}\right] = 0, \tag{21}$$

where:

$$l_{n,\theta_{(m)+1}}(X) = \sum_{n=1}^{N} L_{(m)+1,n-m}(X) \cdot g_{m,n-m}(\theta), \tag{22}$$

$$l_{n,\theta_{(n)+1}}(X) = \sum_{n=1}^{N} L_{m,(n+1)-m}(X) \cdot g_{m,n-m}(\theta). \tag{23}$$

The original functional of Horn and Schunck [12] (12) can be expressed using (21) to define the real-time Hermite optical flow (RT-HOF) as follows

$$E_{RT-HOF}(W) = \int_{\Omega}\left[\left(uL_{0,1}(X) + vL_{1,0}(X) + \frac{\partial L_0(X)}{\partial t}\right)\right.$$

$$+\gamma\sum_{n=1}^{N}\left(ul_{n,\theta_{(m)+1}}(X) + vl_{n,\theta_{(n)+1}}(X) + \frac{\partial l_{n,\theta}(X)}{\partial t}\right) \tag{24}$$

$$\left. + \alpha|\nabla W|^2\right]dX.$$

Using variational calculus, the corresponding Euler-Lagrange equations are:

$$
\left[L_{0,1}^2(X) + \gamma L_{0,1}(X) \sum_{n=1}^{N} \left(l_{n,\theta_{(n)+1}}(X) \right) \right] u
$$
$$
+ \left[L_{0,1}(X)L_{1,0}(X) + \gamma L_{0,1}(X) \sum_{n=1}^{N} \left(l_{n,\theta_{(m)+1}}(X) \right) \right] v \qquad (25)
$$
$$
= \alpha^2 \nabla^2 u - \left[L_{0,1}(X)L_{0t}(X) + \gamma \sum_{n=1}^{N} \left(l_{n,\theta_t}(X) \right) \right],
$$

$$
\left[L_{0,1}(X)L_{1,0}(X) + \gamma L_{1,0}(X) \sum_{n=1}^{N} \left(l_{n,\theta_{(n)+1}}(X) \right) \right] u
$$
$$
+ \left[L_{1,0}^2(X) + \gamma L_{1,0}(X) \sum_{n=1}^{N} \left(l_{n,\theta_{(m)+1}}(X) \right) \right] v \qquad (26)
$$
$$
= \alpha^2 \nabla^2 v - \left[L_{1,0}(X)L_{0t}(X) + \gamma \sum_{n=1}^{N} \left(l_{n,\theta_t}(X) \right) \right],
$$

where $L_{0t}(X) = \frac{\partial L_0(X)}{\partial t}$, $l_{n,\theta_t}(X) = \frac{\partial l_{n,\theta}(X)}{\partial t}$ and $\nabla^2 u$ represents the Laplacian of u.

Finding the solution of (25, 26) for u and v and applying Gauss-seidel iterative method, the final equations hold

$$
u^{n+1} = \bar{u}^n - L_{0,1}(X) \left[\left(L_{0,1}(X) + \gamma \sum_{n=1}^{N} \left(l_{n,\theta_{(n)+1}}(X) \right) \right) \bar{u}^n \right.
$$
$$
+ \left(L_{1,0}(X) + \gamma \sum_{n=1}^{N} \left(l_{n,\theta_{(m)+1}}(X) \right) \right) \bar{v}^n
$$
$$
\left. + L_{0t}(X) + \gamma \sum_{n=1}^{N} \left(l_{n,\theta_t} \right) \right] \Bigg/ \qquad (27)
$$
$$
\left[\alpha^2 + \left(L_{0,1}^2(X) + \gamma L_{0,1}(X) \sum_{n=1}^{N} \left(l_{n,\theta_{(n)+1}}(X) \right) \right) \right.
$$
$$
\left. + \left(L_{1,0}^2(X) + \gamma L_{1,0}(X) \sum_{n=1}^{N} \left(l_{n,\theta_{(m)+1}}(X) \right) \right) \right],
$$

$$v^{n+1} = \bar{v}^n - L_{1,0}(X)\left[\left(L_{0,1}(X) + \gamma \sum_{n=1}^{N}\left(l_{n,\theta_{(n)}+1}(X)\right)\right)\bar{u}^n \right.$$

$$+ \left(L_{1,0}(X) + \gamma \sum_{n=1}^{N}\left(l_{n,\theta_{(m)}+1}(X)\right)\right)\bar{v}^n$$

$$\left. + L_{0t}(X) + \gamma \sum_{n=1}^{N}\left(l_{n,\theta_t}\right)\right] \Bigg/ \tag{28}$$

$$\left[\alpha^2 + \left(L_{0,1}^2(X) + \gamma L_{0,1}(X)\sum_{n=1}^{N}\left(l_{n,\theta_{(n)}+1}(X)\right)\right)\right.$$

$$\left. + \left(L_{1,0}^2(X) + \gamma L_{1,0}(X)\sum_{n=1}^{N}\left(l_{n,\theta_{(m)}+1}(X)\right)\right)\right],$$

where the Laplacian was approximated by $\nabla u \approx k\left(\bar{u}_{i,j} - u_{i,j}\right)$ and the local average was defined by performing a convolutions of u with the kernel

$$\begin{bmatrix} \dfrac{1}{12} & \dfrac{1}{6} & \dfrac{1}{12} \\ \dfrac{1}{6} & 0 & \dfrac{1}{6} \\ \dfrac{1}{12} & \dfrac{1}{6} & \dfrac{1}{12} \end{bmatrix}.$$

To compare the real-time Hermite optical flow with the proposal of Horn and Schunck, both methods were implemented in MATLAB®.

For the Horn and Schunck method, the solution of (12) is given by [12]:

$$u^{n+1} = \bar{u}^n - \left[L_x(X)(L_x(X)\bar{u}^n + L_y(X)\bar{v}^n + L_t(X))\right] \Big/ \left[\alpha^2 + L_x^2(X) + L_y^2(X)\right] \tag{29}$$

$$v^{n+1} = \bar{v}^n - \left[L_y(X)(L_x(X)\bar{u}^n + L_y(X)\bar{v}^n + L_t(X))\right] \Big/ \left[\alpha^2 + L_x^2(X) + L_y^2(X)\right]. \tag{30}$$

In Algorithms 1 and 2, we show the pseudocode for the iterative solution by Gauss-Seidel of the Horn and Schunck and real-time Hermite optical flow methods respectively.

Algorithm 1 Horn and Schunck's optical flow estimation.

1: **procedure** HORN AND SCHUNCK OPTICAL FLOW METHOD
2:     $\alpha \leftarrow$ Smoothing parameter
3:     $Num\_of\_itera \leftarrow$ Number of iterations
4:     $img1 \leftarrow$ Imagen in time t
5:     $img2 \leftarrow$ Imagen in time $t + 1$
6:     $u \leftarrow$ Matrix of zeros of size of $img1$
7:     $v \leftarrow$ Matrix of zeros of size of $img1$
8:     $G \leftarrow$ Square Gaussian filter of size $\sigma/6$
9:     $img1 = img1 \circledast G$ ▷ Smoothing images $img1$ and $img2$
10:     $img2 = img2 \circledast G$
11:     $L_x = \frac{1}{4}\left(img1 \circledast \begin{bmatrix} -1 & 1 \\ -1 & 1 \end{bmatrix} + img2 \circledast \begin{bmatrix} -1 & 1 \\ -1 & 1 \end{bmatrix} \right)$ ▷ Compute spatiotemporal derivatives

12:     $L_y = \frac{1}{4}\left(img1 \circledast \begin{bmatrix} -1 & -1 \\ 1 & 1 \end{bmatrix} + img2 \circledast \begin{bmatrix} -1 & -1 \\ 1 & 1 \end{bmatrix} \right)$

13:     $L_t = \frac{1}{4}\left(img1 \circledast \begin{bmatrix} 1 & 1 \\ 1 & 1 \end{bmatrix} + img2 \circledast \begin{bmatrix} -1 & -1 \\ -1 & -1 \end{bmatrix} \right)$

14:     **for** $i = 1$ to Num_of_itera **do**
15:         $\bar{u} = u \circledast \begin{bmatrix} \frac{1}{12} & \frac{1}{6} & \frac{1}{12} \\ \frac{1}{6} & 0 & \frac{1}{6} \\ \frac{1}{12} & \frac{1}{6} & \frac{1}{12} \end{bmatrix}$ ▷ Compute local averages of the flow

16:         $\bar{v} = v \circledast \begin{bmatrix} \frac{1}{12} & \frac{1}{6} & \frac{1}{12} \\ \frac{1}{6} & 0 & \frac{1}{6} \\ \frac{1}{12} & \frac{1}{6} & \frac{1}{12} \end{bmatrix}$

17:         Update u and v using the Eq. (29) and (30)
18:     **end for**
19:     **return** u, v
20: **end procedure**

Algorithm 2 Real-time Hermite optical flow estimation.

1: **procedure** REAL-TIME HERMITE OPTICAL FLOW METHOD
2:    $\alpha \leftarrow$ Smoothing parameter
3:    $\gamma \leftarrow$ Weight parameter
4:    *Num_of_itera* $\leftarrow$ Number of iterations
5:    $N \leftarrow$ Maximum Hermite polynomial degree
6:    $M \leftarrow$ Square Gaussian window size $(M+1)$
7:    *img1* $\leftarrow$ Imagen in time t
8:    *img2* $\leftarrow$ Imagen in time $t+1$
9:    $u \leftarrow$ Matrix of zeros of size of *img1*
10:   $v \leftarrow$ Matrix of zeros of size of *img1*
11:   Compute the analysis Hermite filters $D_{m,n-m}$ from Eq. (5) and (6):
12:   **for** $n = 0$ to N **do**
13:      **for** $m = 0$ to n **do**
14:         $L_{m,n-m} = img1 \circledast D_{m,n-m}$ $\triangleright$ Cartesian Hermite coefficients *img1*
15:         $Lw_{m,n-m} = img2 \circledast D_{m,n-m}$ $\triangleright$ Cartesian Hermite coefficients *img2*
16:         $\theta = \arctan \frac{L_{0,1}}{L_{1,0}}$ $\triangleright$ Phase of the gradient of Hermite coefficients *img1*
17:         $\theta w = \arctan \frac{Lw_{0,1}}{Lw_{1,0}}$ $\triangleright$ Phase of the gradient of Hermite coefficients *img2*
18:         $g_{m,n-m}(\theta) = \sqrt{\binom{n}{m}} \cos^m \theta \cdot \sin^{n-m} \theta$ $\triangleright$ Angle functions *img1*
19:         $gw_{m,n-m}(\theta w) = \sqrt{\binom{n}{m}} \cos^m \theta w \cdot \sin^{n-m} \theta w$ $\triangleright$ Angle functions *img2*
20:         $l_{n,\theta} = \sum \sum_{n=1}^{N} \left(L_{m,n-m} \right) \left(g_{m,n-m}(\theta) \right)$ $\triangleright$ ST coefficients *img1*
21:         $l_{n,\theta_{(m)+1}}(X) = \sum_{n=1}^{N} L_{(m)+1,n-m}(X) \cdot g_{m,n-m}(\theta)$
22:         $l_{n,\theta_{(n)+1}}(X) = \sum_{n=1}^{N} L_{m,(n+1)-m}(X) \cdot g_{m,n-m}(\theta)$
23:         $lw_{n,\theta} = \sum \sum_{n=1}^{N} \left(Lw_{m,n-m} \right) \left(gw_{m,n-m}(\theta) \right)$ $\triangleright$ ST coefficients *img2*
24:      **end for**
25:   **end for**
26:   $L_{0t} = Lw_{0,0} - L_{0,0}$ $\triangleright$ Temporal derivatives
27:   $l_{n,\theta_t} = \sum_{n=1}^{N} lw_{n,\theta w} - \sum_{n=1}^{N} l_{n,\theta}$
28:   **for** $i = 1$ to Num_of_itera **do**
29:      $\bar{u} = u \circledast \begin{bmatrix} \frac{1}{12} & \frac{1}{6} & \frac{1}{12} \\ \frac{1}{6} & 0 & \frac{1}{6} \\ \frac{1}{12} & \frac{1}{6} & \frac{1}{12} \end{bmatrix}$ $\triangleright$ Compute local averages of the flow
30:      $\bar{v} = v \circledast \begin{bmatrix} \frac{1}{12} & \frac{1}{6} & \frac{1}{12} \\ \frac{1}{6} & 0 & \frac{1}{6} \\ \frac{1}{12} & \frac{1}{6} & \frac{1}{12} \end{bmatrix}$
31:      Update u and v using the Eq. 27 and 28
32:   **end for**
33:   **return** u, v
34: **end procedure**

---

5 Implementation on Navigation Mobile Robots

This section describes a case study in which a mobile robot implements a movement detection controller based on the Hermitian optical flow approach presented earlier. For experimental purposes, this controller was developed and implemented in a LEGO® robot with a mounted webcam, and the whole system communicates with MATLAB which computes the control system. In that sense, the whole robotic system is introduced in this section as well as the design of the control law.

5.1 Description of the System

In this case study, a mobile robot was implemented with the LEGO® Mindstorms EV3 as both the mechanical and the electrical platform because it is easy and practical to use. In particular, the mobile robot was built as in a tank configuration using two direct current (DC) motors and two rubber bands as actuators, and the robot was planned to be in a differential steering configuration. In addition, a webcam was mounted on the robot to be used as the sensor. For experimental purposes, the intelligent brick of the LEGO® Mindstorms EV3 set was used as an interface between the mobile robot and a computer with MATLAB®, as the latter was employed to compute the control of the whole system. Both the webcam and the intelligent brick communicated with the computer via USB. Figure 7 shows a diagram of the robotic system. The technical specifications of the robotic system are summarized in Table 1.

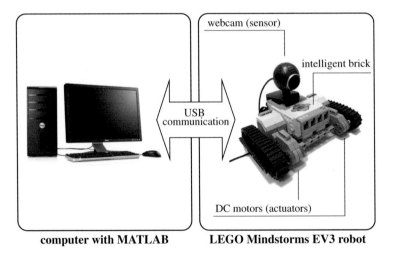

computer with MATLAB   **LEGO Mindstorms EV3 robot**

Fig. 7 Diagram of the robotic system employed in the case study

Table 1 Technical specifications of the robotic system

Component	Description
Webcam sensor	1.3 M pixels, 1280 × 1024 max resolution
DC motor actuators	9 V-input, 0.55 A nominal, 2.03 W mechanical power, 4.95 W electrical power, 160 rpm max nominal
USB communication	9600 baud rate
Windows-based workstation	Intel® Xeon® six-core processor 2.6 GHz, 32 GB RAM

5.2 Characterization of Hermite Optical Flow Method in Robot Navigation

In order to develop a nature-inspired control law for robot navigation, a characterization of the real-time Hermite optical flow method was performed. This characterization considered a description of the performance of the RT-HOF in terms of the displacement values measured between two images. Two experiments were conducted: a displacement measurement using RT-HOF without timeout limitations between frames (soft real-time system), and the second experiment measures the displacement values using RT-HOF with hard timeout deadlines between frames (hard real-time system).

5.2.1 Characterization in a Soft Real-Time Hermite Optical Flow System

The first experiment for characterizing the proposed real-time Hermite optical flow algorithm was developed under a soft real-time framework avoiding any deadline times, i.e., the software is free to use unlimited time for computing displacements between two images. In the experiment, the LEGO® robot was located statically in front of a reconfigurable wall. That wall worked as the background of the scene, with four possible configurations: white, dry-leaves, blue-square, and red-stripes backgrounds, as shown in Fig. 8. Then, another LEGO® robot started to move in front of the main robot with a certain speed (also configurable in three possible states: low, medium and hard). A short video was recorded using the same webcam mounted on the main robot.

For analysis purposes, a test point marked in red color was previously located at the middle of the secondary robot. Then, manual segmentation was done to find the test point in every frame of the short video. After that, the RT-HOF algorithm measured the displacement of the test point in the whole video. Figure 9 shows the mean of magnitude and phase of the displacement at the test point for some representative cases, best and worst results, of the secondary robot using all backgrounds. In Tables 2 and 3, we show the results in magnitude and phase of the

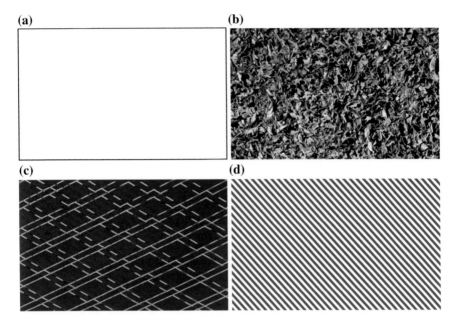

Fig. 8 Examples of backgrounds used in the experiments: **a** white, **b** dry leaves, **c** blue squares, **d** red stripes

vector of displacement at the test point for each speed case and for each background. Notice that the displacement of the test point relatively measures the displacement of the whole secondary robot (e.g., as a dynamic obstacle). It is evident from Fig. 9 that the RT-HOF method can distinguish different levels of speed in mobile obstacles in both plain and texturized backgrounds without any difficulties. For comparison purposes, the Horn method was also computed as seen in Fig. 9.

5.2.2 Characterization in a Hard Real-Time Hermite Optical Flow System

The second experiment for characterizing the proposed RT-HOF method was developed under a hard real-time framework with a time rate of 100 ms. The same scene was occupied with four different backgrounds (see Fig. 8) and three levels of speed in the dynamic obstacle. However, an automated segmentation was employed to extract only the measured displacements of the dynamic obstacle online (secondary robot), and then the mean displacement (magnitude and angle) was obtained for each pair of frames in the webcam streaming. Figure 10 shows some the mean displacement at the mobile obstacle for each speed case using all background configurations, and Tables 4 and 5 summarizes the whole experimental results.

Some conclusions can be drawn from Figs. 9 and 10, and Tables 4 and 5. First, the Horn-based method generates a greater angle than Hermite-based method.

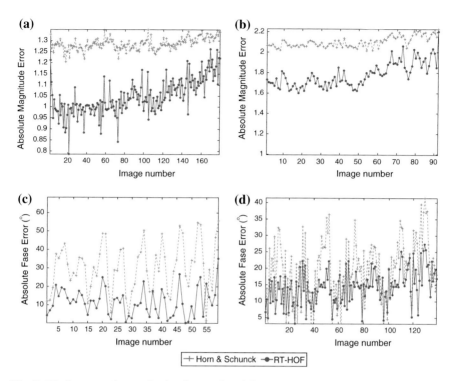

Fig. 9 Displacement characterization for a soft real-time system, best and worst results for the mean of magnitude and phase: **a** best absolute magnitude error with red stripes background at speed one, **b** worst absolute magnitude error with dry leaves background at speed two, **c** best absolute phase error with blue squares background at speed three, **d** worst absolute phase error with squares background at speed three

Table 2 Displacement characterization in magnitude for a soft real-time system

Background	Method	Mean magnitude error $\mu \pm \sigma$ [px]		
		m_1	m_2	m_3
White	Horn	1.34 ± 0.02	1.94 ± 0.44	2.16 ± 0.63
	Hermite	1.10 ± 0.07	1.64 ± 0.36	1.86 ± 0.55
Dry-leaves	Horn	1.28 ± 0.03	1.70 ± 0.41	1.85 ± 0.58
	Hermite	1.09 ± 0.07	1.46 ± 0.33	1.66 ± 0.54
Blue squares	Horn	1.18 ± 0.04	1.62 ± 0.38	1.78 ± 0.53
	Hermite	1.00 ± 0.09	1.34 ± 0.28	1.55 ± 0.47
Red stripes	Horn	1.90 ± 0.56	1.75 ± 0.41	1.90 ± 0.56
	Hermite	1.04 ± 0.08	1.42 ± 0.32	1.60 ± 0.48

While Horn-based method can distinguish large speeds in both white and blue-squares backgrounds, it cannot recognize large speeds in both dry-leaves and red-stripes backgrounds significantly. In addition, Hermite-based method can

Table 3 Displacement characterization in phase for a soft real-time system

Background	Method	Mean angle error $\mu \pm \sigma$ [°]		
		θ_1	θ_2	θ_3
White	Horn	19.61 ± 9.31	30.79 ± 14.45	34.79 ± 15.53
	Hermite	14.39 ± 5.24	13.09 ± 6.70	14.07 ± 7.33
Dry-leaves	Horn	16.61 ± 9.94	30.17 ± 19.23	32.23 ± 20.91
	Hermite	15.55 ± 5.38	17.98 ± 7.47	18.57 ± 10.00
Blue squares	Horn	14.85 ± 16.17	22.93 ± 15.04	24.82 ± 16.05
	Hermite	16.17 ± 5.87	15.26 ± 6.90	14.57 ± 7.84
Red stripes	Horn	15.25 ± 9.94	25.75 ± 16.04	28.32 ± 2.11
	Hermite	13.81 ± 5.23	13.90 ± 6.44	13.82 ± 7.526

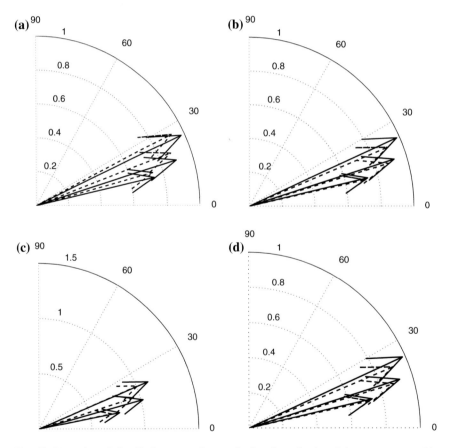

Fig. 10 Examples of the displacement characterization for a hard real-time system: **a** white background, **b** dry-leaves background, **c** blue squares background, **d** red stripes background. (*strong-line*) Hermite-based method, (*dashed-line*) Horn-based method. *Arrows* represent vectors of displacement

Table 4 Displacement characterization in magnitude for a hard real-time system

Background	Method	Magnitude $\mu \pm \sigma$ [px/s]		
		m_1	m_2	m_3
White	Horn	0.74 ± 0.09	0.88 ± 0.08	0.93 ± 0.06
	Hermite	0.74 ± 0.07	0.90 ± 0.09	0.98 ± 0.05
Dry-leaves	Horn	0.72 ± 0.063	0.88 ± 0.05	0.91 ± 0.07
	Hermite	0.74 ± 0.062	0.92 ± 0.04	0.98 ± 0.05
Blue squares	Horn	0.70 ± 0.06	0.94 ± 0.10	1.01 ± 0.05
	Hermite	0.78 ± 0.07	1.02 ± 0.04	1.11 ± 0.02
Red stripes	Horn	0.69 ± 0.07	0.86 ± 0.08	0.92 ± 0.040
	Hermite	0.74 ± 0.08	0.90 ± 0.11	1.00 ± 0.042

Table 5 Displacement characterization in phase for a hard real-time system

Background	Method	Angle $\mu \pm \sigma$ [°]		
		θ_1	θ_2	θ_3
White	Horn	15.10 ± 3.19	20.56 ± 4.27	26.90 ± 5.21
	Hermite	12.84 ± 1.89	17.33 ± 3.96	25.02 ± 4.43
Dry-leaves	Horn	12.38 ± 3.20	16.60 ± 3.47	22.47 ± 6.56
	Hermite	13.04 ± 2.34	17.48 ± 3.09	24.23 ± 2.38
Blue squares	Horn	11.22 ± 2.82	17.47 ± 4.16	22.65 ± 4.38
	Hermite	10.99 ± 2.11	15.05 ± 1.94	22.47 ± 4.08
Red stripes	Horn	15.33 ± 3.57	20.68 ± 2.97	27.41 ± 4.96
	Hermite	14.87 ± 2.66	19.06 ± 2.48	27.54 ± 3.59

distinguish between speeds in any background. The overall results of these experiments are that real-time Hermite optical flow can be used as a nature-inspired control law because its performance can distinguish between different levels of speeds in dynamic obstacles, and it responds well in hard real-time framework.

5.3 Design of the Nature-Inspired Control Law

The LEGO® robot was controlled using a nature-inspired control law, based on the real-time Hermite optical flow method. Following, we describe the methodology to design the nature-inspired controller.

This work adopted the agent-environment cycle methodology for designing the control law in the robot, as shown in Fig. 11. This aims that the robot perceives the state of the environment (e.g., the scene in which the robot is interacting), then a specific robot architecture defines how this perception is interpreted to finally get a decision of the type of action it will perform in the environment. Once the action is selected, the robot executes it and the agent-environment cycle assumes that the

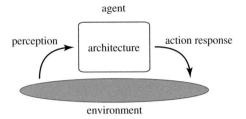

Fig. 11 Agent-environment cycle

state of the environment changes, giving the opportunity to the robot to start the process again.

Thus, the design of the nature-inspired control law in this work refers to find a suitable robot architecture. In Fig. 12 the final design of the robot architecture used in this case study is depicted. As shown, the robot starts the process acquiring images at 100 ms of frame rate. Then, a pair of two adjacent images are analyzed using the real-time Hermite optical flow. A map of the displacements is then obtained showing where objects in images are moving. Next, an automated segmentation is done in order to get only the displacements that reaches an empirical threshold, assuming that dynamic obstacles perform displacements greater than this

Fig. 12 Diagram of the nature-inspired controller used in the case study

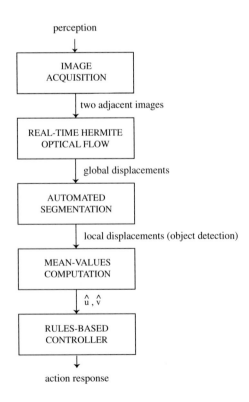

Fig. 13 An example of the field observed by the robot

threshold. For this case study, a threshold value of $T_{seg} = 0.5$ was selected (using a trial-and-error approach). In addition, this case study supposes that there is one dynamic obstacle in front of the LEGO® robot at most. Thus, the automated segmentation finds the dynamic obstacle and computes its local displacements.

In Fig. 13, we show an example of the field of displacements segmented observed by the LEGO® robot when it moves and other robot approximates.

Later on, the robot computes the mean value $\hat{u}$ of the horizontal component of displacements and the mean value $\hat{v}$ of the vertical component of the displacements. Also, the resultant angle $\hat{\theta}$ of the displacements is computed. All these values give to the robot some advice about the movement of the dynamic object in the two dimensions projected on the image. Then, a rule-based controller was designed in terms of the mean values $\hat{u}$ and $\hat{v}$ and the angle $\hat{\theta}$. Algorithm 3 shows the prototype of the rules-based controller. As noted, it requires two thresholds T_u and T_v associated to the horizontal and vertical components of the dynamic obstacle's displacement.

Algorithm 3 Rules–based controller used in the case study.

1: **if** $\|\hat{u}\| > T_u$ **and** $\|\hat{v}\| > T_v$ **then**
2:     **if** $\|\hat{\theta}\| > T_\theta$ **then**
3:         braking or steering–to–the–right
4:     **else if** $\|\hat{\theta}\| < T_\theta$ **and** $\|\hat{\theta}\|! = 0$ **then**
5:         braking or steering–to–the–left
6:     **else**
7:         go–forward
8:     **end if**
9: **else**
10:     go–forward
11: **end if**

Two experiments were conducted to find the thresholds T_u and T_v for the rules-based controller. The first experiment consisted on having a fixed obstacle in front of the LEGO® robot, and this one moves straight until reached the fixed

256   E. Moya-Albor et al.

obstacle. The webcam mounted in the robot acquired images at 100 ms of frame
rate. The real-time Hermite optical flow with automated segmentation obtained the
mean values $\hat{u}$ and $\hat{v}$ during the translation. Then both mean values were plotted as
shown in Fig. 14. As noted, when the robot is far from the obstacle the $\hat{v}$ value is
small, when the robot is close to the obstacle the $\hat{v}$ value is large (slope of 0.02
px/s). The $\hat{u}$ value is not affected (slope of $7.1 \times 10 - 4$ px/s). The same experiment
was repeated but now the fixed obstacle was replaced with a dynamic obstacle with
constant speed. Results of this experiment are shown in Fig. 15. Notice that when $\hat{u}$
is minimum and $\hat{v}$ is maximum, the mobile obstacle is placed in front of the robot.
From the latter figures it can be seen that the most appropriate threshold values are
$T_u = 0.6$ px and $T_v = 0.5$ px.

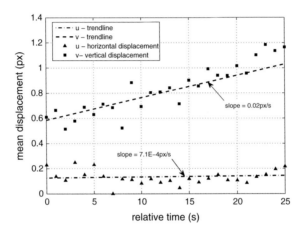

Fig. 14 Experiment 1 for extracting the threshold values of the components in displacements

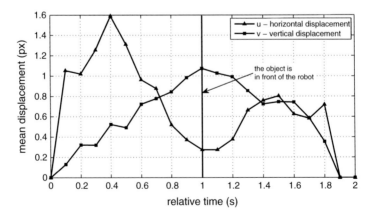

Fig. 15 Experiment 2 for extracting the threshold values of the components in displacements

In addition, a threshold T_θ is needed for determining the direction of the mobile obstacle. In this case study, a value of $T_\theta = 40°$ was used (obtained by trial-and-error). At last, the robot takes a decision from the rules-based controller, picking the more suitable action. For this case study three possible actions were designed: *go-forward*, *braking*, or *steering*. Once the LEGO® robot performs the action, the whole process is repeated again.

5.4 Experimental Results for Dynamic Obstacles Avoidance

This case study implemented the proposed real-time Hermite optical flow method as a nature-inspired controller for dynamic obstacles avoidance in robots (Fig. 12). In particular, a LEGO® robot with a webcam, as the unique sensor, was employed. Two experiments were performed as follows: a braking action and a steering action.

5.4.1 Braking Action

This experiment consisted on braking the LEGO® robot when a mobile obstacle is presented in front of it. Once the obstacle moves on, the robot can go straight again. Figure 16 show the trajectories of the robot in that situation with three possible constant speeds of the mobile obstacle: *low*, *medium* and *high*. It is evident from Fig. 16 that in all cases the LEGO® robot can stop without colliding with the obstacle.

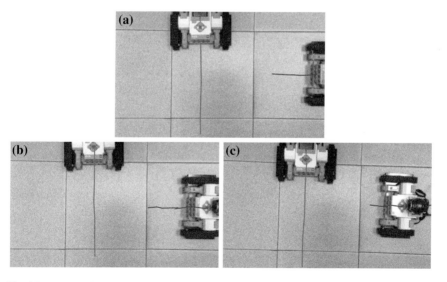

Fig. 16 An example of the trajectories and evidence of the braking actions in the robot, using a mobile obstacle with **a** *low*, **b** *medium* and **c** *high* speed

In Fig. 17 we show the image sequence of the trajectory followed for the LEGO® robot for the *medium* speed, where the braking action is observed in the three images of the central row.

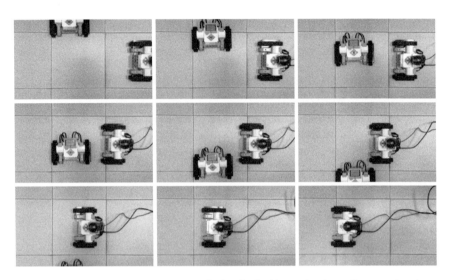

Fig. 17 Image sequence of the trajectory of robot for braking action in *medium* speed

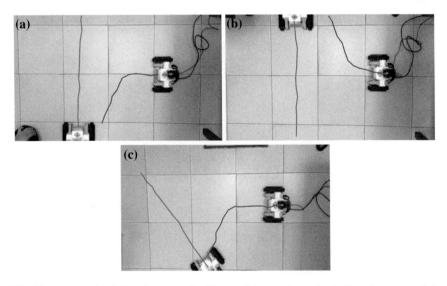

Fig. 18 An example of the trajectory and evidence of the steering action in the robot: **a** turns left, **b** turns right and **c** avoidance behavior when a mobile obstacle moves diagonally

5.4.2 Steering Action

This experiment consisted on steering the LEGO® robot when a mobile obstacle is presented in front of it. In particular, the robot steers in the opposite direction of the obstacle. The trajectory obtained from the experiment when the robot turns left and when it turns right are shown in Fig. 18a–b, and when it avoids a mobile obstacle

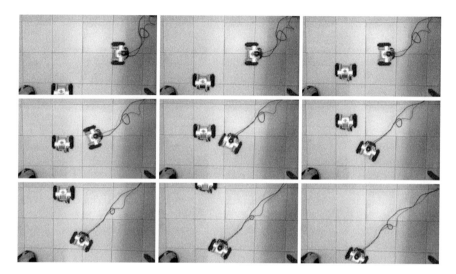

Fig. 19 Image sequence of the trajectory of robot for the left turn action

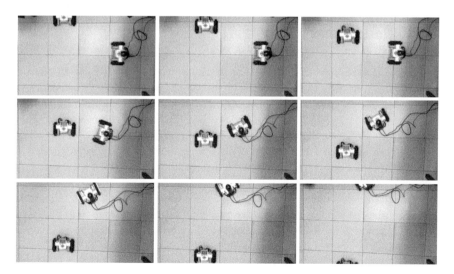

Fig. 20 Image sequence of the trajectory of robot for the right turn action

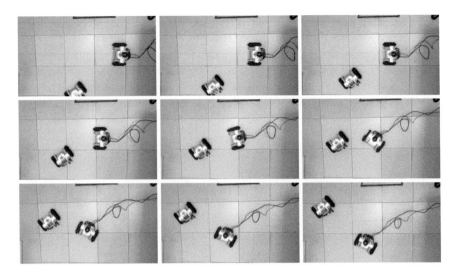

Fig. 21 Image sequence of the path followed for the robot when a mobile obstacle moves diagonally

that moves diagonal is shown in Fig. 18c. Notice that the LEGO® robot can avoid mobile obstacles in all situations.

The paths followed for the LEGO® robot are shown in the image sequences of Figs. 19 and 20 respectively, where the turn's choice of robot in both cases is opposite to direction of the obstacle in motion. In Fig. 21 we show the image sequence for the avoidance behavior of a mobile obstacle that moves diagonal.

6   Conclusions and Future Work

This work presented and described a new Hermite optical flow (RT-HOF) method for real-time purposes as a nature-inspired technique for computer vision. As described, this new real-time Hermite optical flow method is easier to implement and faster than the original Hermite optical flow method. In particular, this work focuses on using that approach for controlling a mobile robot.

As a case study, a LEGO robot with a webcam was occupied to design a controller for mobile obstacles avoidance. A detailed description of the design methodology of the controller was included, from an agent-environment based model to the robot's architecture. Different tests were run for characterizing and validating the controller. In fact, several comparisons between the proposed RT-HOF method with Horn and Schunck-based algorithm were developed in terms of background textures and velocities of mobile obstacles, in which RT-HOF method resulted to be very effective for real-time purposes. On the other hand, the design of the nature-inspired controller was described such that the robot can perform actions based on its perception of the environment. Some guidelines were also presented to find empirical parameters in the controller.

Results confirm that the nature-inspired controller based on the new real-time Hermite optical flow method can avoid mobile obstacles in two different approaches: braking and steering when mobile obstacles are close to the robot. In addition, background textures and colours did not affect the performance of the controller. To this end, it is remarkable to say that this nature-inspired approach allows robots avoid mobile obstacles even though the parameters of the whole robotics system are unknown. In contrast with other works that perform obstacle avoidance with optical flow methods, the proposed RT-HOF: (i) is fast enough to compute an approximate solution of displacements between images that can be used as a visual perception in robotic systems, (ii) is robust to different background textures and colours, (iii) allows designing a nature-inspired controller, and (iv) its performance does not depend on parameters of a robotic system.

In future work, other improvements to the real-time Hermite optical flow method will be attended to increase the accuracy in the approximation of displacements and multi-resolution approaches will also be considered. Hardware independence is also considered for autonomous navigation of robots. In addition, other environmental characteristics will be evaluated, e.g. variation of light intensity. Also, the set of actions in the robot will be increased to perform better obstacle avoidance. Lastly, applications of this approach in the real-world will be investigated.

References

1. Abdagic, A., Tanovic, O., Aksamovic, A., Huseinbegovic, S.: Counting traffic using optical flow algorithm on video footage of a complex crossroad. In: ELMAR, 2010 PROCEEDINGS, pp. 41–45 (2010)
2. Barron, J.L., Fleet, D.J., Beauchemin, S.S.: Performance of optical flow techniques. Int. J. Comput. Vis. **12**(1), 43–77 (1994). doi:10.1007/BF01420984, http://dx.doi.org/10.1007/BF01420984
3. Beauchemin, S.S., Barron, J.L.: The computation of optical flow. ACM Comput. Surv. **27**(3), 433–467 (1996)
4. Bloom, J., Reed, T.: A gaussian derivative-based transform. IEEE Trans. Image Process. **5**(3), 551–553 (1996)
5. Bruhn, A., Weickert, J., Schnörr, C.: Combining the advantages of local and global optic flow methods. In: Proceedings of the Twenty-fourth DAGM Symposium on Pattern Recognition, Springer, London, UK, pp. 454–462 (2002)
6. Bruhn, A., Weickert, J., Schnörr, C.: Lucas/Kanade meets horn/Schunck: combining local and global optic flow methods. Int. J. Comput. Vis. **61**(3), 211–231 (2005)
7. Duhamel, P.E., Perez-Arancibia, N., Barrows, G., Wood, R.: Biologically inspired optical-flow sensing for altitude control of flapping-wing microrobots. Mechatron. IEEE/ASME Trans. **18**(2), 556–568 (2013). doi:10.1109/TMECH.2012.2225635
8. Galvin, B., Mccane, B., Novins, K., Mason, D., Mills, S.: Recovering motion fields: an evaluation of eight optical flow algorithms. In: British Machine Vision Conference, pp. 195–204 (1998)
9. Gibson, J.J.: The perception of the visual world. Am. J. Psychol. **64**, 622–625 (1951)

10. Henry Guennadi Levkin.: Imageprocessing/videocodecs/programming. http://www.hlevkin. com/TestImages/additional.htm (2004). Accessed 20 Jan 2013
11. Honegger, D., Greisen, P., Meier, L., Tanskanen, P., Pollefeys, M.: Real-time velocity estimation based on optical flow and disparity matching. In: 2012 IEEE/RSJ International Conference on Intelligent Robots and Systems (IROS), pp. 5177–5182 (2012). doi:10.1109/ IROS.2012.6385530
12. Horn, B.K.P., Schunck, B.G.: Determining optical flow. Artif. Intell. 17(1–3), 185–203 (1981)
13. Koenderink, J., van Doorn, A.: Generic neighborhood operators. IEEE Trans. Pattern Anal. Mach. Intell. 14(6), 597–605 (1992)
14. Lan, H., Mei, S.J., Hua, C.P., Hua, C.G.: Visual navigation for UAV using optical flow estimation. In: Control Conference (CCC), 2014 33rd Chinese, pp. 816–821 (2014). doi:10. 1109/ChiCC.2014.6896732
15. Liu, C., Freeman, W.T., Adelson, E.H., Weiss, Y.: Human-assisted motion annotation. http:// people.csail.mit.edu/celiu/motionAnnotation/database/cameramotion.zip (2008). Accessed 12 Oct 2011
16. Lucas, B.D., Kanade, T.: An iterative image registration technique with an application to stereo vision. In: Proceedings of the Seventh International Joint Conference on Artificial Intelligence (IJCAI '81), pp. 674–679 (1981)
17. Marr, D., Hildreth, E.: A theory of edge detection. R. Soc. Lond. B 207, 187–217 (1980)
18. Martens, J.B.: The hermite transform-theory. IEEE Trans. Acoust. Speech Signal Process. 38 (9), 1595–1606 (1990)
19. Motokucho, T., Oda, N.: Vision-based human-following control using optical flow field for power assisted wheelchair. In: 2014 IEEE 13th International Workshop onAdvanced Motion Control (AMC), pp. 266–271 (2014). doi:10.1109/AMC.2014.6823293
20. Moya-Albor, E.: Optical flow estimation using the hermite transform. Ph.D. thesis, Universidad Nacional Autónoma de México, México City, México (2013)
21. Moya-Albor, E., Escalante-Ramírez, B., Vallejo, E.: Optical flow estimation in cardiac CT images using the steered Hermite transform. Signal Process. Image Commun. 28(3), 267–291 (2013a). doi:10.1016/j.image.2012.11.005, http://www.sciencedirect.com/science/article/ pii/S092359651200207X
22. Moya-Albor, E., Escalante-Ramírez, B., Vallejo, E.: Optical flow estimation of the heart's short axis view using a perceptual approach, vol. 8922, pp. 892,206–892,206–12 (2013b). doi:10.1117/12.2041960, http://dx.doi.org/10.1117/12.2041960
23. Nagel, H.H.: Constraints for the estimation of displacement vector fields from image sequences. In: Proceedings of the Eighth International Joint Conference on Artificial Intelligence, vol. 2, pp. 945–951 (1983)
24. Nagel, H.H., Enkelmann, W.: An investigation of smoothness constraints for the estimation of displacement vector fields from image sequences. IEEE Trans. Pattern Anal. Mach. Intell. 8 (5), 565–593 (1986)
25. Papenberg, N., Bruhn, A., Brox, T., Didas, S., Weickert, J.: Highly accurate optic flow computation with theoretically justified warping. Int. J. Comput. Vis. 67(2), 141–158 (2006)
26. Sakitt, B., Barlow, H.B.: A model for the economical encoding of the visual image in cerebral cortex. Biol. Cybern. 43(2), 97–108 (1982)
27. Serra, P., Cunha, R., Hamel, T., Silvestre, C., Le Bras, F.: Nonlinear image-based visual servo controller for the flare maneuver of fixed-wing aircraft using optical flow. IEEE Trans. Control Syst. Technol. 23(2), 570–583 (2015). doi:10.1109/TCST.2014.2330996
28. Sharma, R., Taubel, G., Yang, J.S.: An optical flow and hough transform based approach to a lane departure warning system. In: 11th IEEE International Conference on Control Automation (ICCA), pp 688–693 (2014). doi:10.1109/ICCA.2014.6871003
29. Treisman, A.: Perceptual grouping and attention in visual search for features and for objects. J. Exp. Psychol. Hum. Percept. Perform. 8(2), 194 (1982)
30. Valette, F., Ruffier, F., Viollet, S., Seidl, T.: Biomimetic optic flow sensing applied to a lunar landing scenario. In: 2010 IEEE International Conference on Robotics and Automation (ICRA), pp. 2253–2260 (2010). doi:10.1109/ROBOT.2010.5509364

31. van Dijk, A.M., Martens, J.B.: Image representation and compression with steered Hermite transforms. Sig. Process. **56**(1), 1–16 (1997)
32. Wang, Z., Zhao, J.: Optical flow based plane detection for mobile robot navigation. In: 2011 9th World Congress on Intelligent Control and Automation (WCICA), pp. 1156–1160 (2011). doi:10.1109/WCICA.2011.5970697
33. Wilson, R., Granlund, G.H.: The uncertainty principle in image processing. IEEE Trans. Pattern Anal. Mach. Intell. PAMI. **6**(6), 758–767 (1984)
34. Young, R.A.: The Gaussian derivative theory of spatial vision: analysis of cortical cell receptive field line–weighting profiles. Tech. Rep. GMR-4920, General Motors Research Laboratories, Detroit, Mich, USA (1985)
35. Young, R.A.: Simulation of human retinal function with the Gaussian derivative model. In: Proceedings of the IEEE International Conference on Computer Vision and Pattern Recognition, pp. 564–569 (1986)
36. Young, R.A.: The Gaussian derivative model for spatial vision: I. Retinal Mech. Spat. Vis. **2**(4), 273–293 (1987)
37. Zamudio, Z., Lozano, R., Torres, J., Campos, E.: Stabilization of a helicopter using optical flow. In: 2011 8th International Conference on Electrical Engineering Computing Science and Automatic Control (CCE), pp. 1–6 (2011). doi:10.1109/ICEEE.2011.6106690

Evolutionary Function Approximation
for Gait Generation on Legged Robots

Oscar A. Silva and Miguel A. Solis

Abstract Reinforcement learning methods can be computationally expensive. Their cost is prone to be higher when the cardinality of the state space representation becomes larger. This curse of dimensionality plays an important role on our work, since gait generation by using more degrees of freedom at each leg, implies a bigger state space after discretization, and look-up tables become impractical. Thus, appropriate function approximators are needed for such kind of tasks on robotics. This chapter shows the advantage of using reinforcement learning, specifically within the batch framework. A neuroevolution of augmenting topologies scheme is used as function approximator, a particular case of a topology and weight evolving artificial neural network which has proved to outperform a fixed-topology network for certain tasks. A comparison between function approximators within the batch reinforcement learning approach is tested on a simulated version of an hexapod robot designed and already built at our undergraduate and graduate students group.

Keywords Artificial neural networks · Reinforcement learning · Robotics

1 Natural Inspiration of the Work

We present an hexapod robot called Crabot, which is biologically inspired by a crab. The main objective of this work is to generate a walking pattern for this robot, using experiences of its interactions with the environment. The methods presented in this work, lie inside the reinforcement learning area, a particular type of machine learning with its roots on psychology. Reinforcement learning consists on obtaining

O.A. Silva
Innovación y Robótica Estudiantil UTFSM, Valparaíso, Chile
e-mail: oscar.silvam@alumnos.usm.cl

M.A. Solis (✉)
Centro de Robótica UTFSM, Valparaíso, Chile
e-mail: miguel.solis@alumnos.usm.cl

© Springer International Publishing Switzerland 2016
H.E. Ponce Espinosa (ed.), *Nature-Inspired Computing for Control Systems*,
Studies in Systems, Decision and Control 40, DOI 10.1007/978-3-319-26230-7_10

a desired behavior by allowing the environment to provide reinforcement signals to the agent, quantifying the impact of choosing a particular action.

2 Introduction

When we talk about reinforcement learning (RL), we refer to the area of machine learning, concerned about solving sequential decision problems modelled by markovian decision processes (MDPs) [39]. Nevertheless, the applications spectrum has been extended to areas such as robotics [22] or control theory [14, 26, 40].

In psychology, in a similar form to the popular theory of left and right halves of the brain, the study of top and bottom parts of the brain (see [24]) sets that the top part is responsible of planification and execution of decisions, while the bottom part is in charge of classification and representation of data. Although RL has its roots in studies of animal learning, to the best knowledge of the authors, there is not evidence of a formal derivation of RL from this particular psychological study. Nevertheless, the findings of the study are similar to how RL solves sequential decision problems, given a suitable state representation, which classifies environment and agent information. Then, based on this data, the agent has to choose the best possible action, by typically using state-action value function, Q, the expected sum of reinforcement signals starting from the current state. This signal is an abstraction of how desirable was executing that action for that particular state.

Classical RL methods, like Q-learning [42], assume that the problem is given in terms of a discrete state space representation, requiring to store data from state-action value function, Q, in look-up tables. As the intuition may suggest, these tabular methods become impractical as the state space becomes larger, or impossible to use in the case of having a continuous state space representation. Thus, function approximators are needed.

Within the control framework, RL has been used in tracking problems, whose objective is to make the output of the process being controlled, to follow a reference trajectory. For instance, [31] tackles the linear quadratic tracking (LQT) problem, where the process to be controlled is modeled by a linear system, and the performance index is quadratic in terms of the state. Reference [21] goes further on this LQT problem, with optimality constraints, where the performance index has to be minimized, so the best performance that the controller is capable to generate, has to be the best performance that any controller could ever generate.

When solving adaptive control problems with RL methods, the user can benefit by avoiding the generation of models for processes with high-complexity dynamics. Vast majority of RL methods can be grouped into algorithms based on the Q-function, and actor-critic algorithms. In the latter, the critic is responsible for estimating the long-term utility for each state, and the actor is the subsystem that chooses an optimal action in each state. In this context, [41] obtains an online actor-critic algorithm for optimal control, where the performance index is not necessarily dependent on the state of the process. On the other hand, [20] develops an algorithm based on the action-value function, Q, for the LQT problem when the process to be controlled has unknown dynamics.

Batch reinforcement learning (BRL) does not attempt to tackle this curse of dimensionality, but aims to use transition samples in a more efficient way. It collects a bunch of experiences, and then uses them to update action influences instead of updating the action value function in an incremental way [25].

The batch framework includes algorithms like experience replay (ER) or fitted Q iteration (FQI), where the latter has been taken as basis for derivations like Neural FQI [36] or FQI with alternative update rules like Q-Batch [4].

Both RL and BRL approaches have been used for gait generation on robotics [33, 46], where due to the number of degrees of freedom per leg, the state space becomes large enough for having to introduce function approximation. Artificial neural networks (ANN) are commonly used for such approximation purposes [10].

Neuroevolution of augmenting topologies (NEAT) [38] has been successfully applied on RL tasks, by typically evolving a population of action selectors, i.e., this structure has one input for each state feature and one output for each possible action. Nevertheless, it also has been used on the classical RL domain, being combined with Q-learning for giving place to NEAT+Q [43]. This method has proved to outperform the standard (fixed-topology) neural networks used for function approximation, but to the best of the knowledge of the authors, this setup in which the neural network is allowed to adjust its topology along with its weights, has not been found in literature for the batch case. Thus, this chapter introduce and evaluates the performance of neuroevolutive fitted q iteration (NEFQI), for a walking pattern generation on legged robots.

NEFQI makes use of NEAT for function approximation within the BRL framework, making an efficient use of collected data, regardless the chosen topology for the neural network, since it would start with no hidden layers and then its structure will evolve accordingly. Note that the fitness of the current network will be evaluated just the last time the whole batch is visited, in order to ensure that the different networks are compared once all the weights have been adjusted after the action-value samples have been propagated.

NEAT corresponds to one of the most prominent topology and weight evolving artificial neural networks (TWEANN) algorithms. Its (genetic) encoding system is such that each individual represents an ANN, also known as a genome.

In general, the initial population on a TWEANN is a collection of random topologies, which is problematic given that it could be possible that no path between a given input to an output is found. In order to evolve minimal solutions, NEAT starts with minimal topology, i.e., there are no hidden nodes, and inputs are directly connected to outputs. Forthcoming structures are obtained by mutation, stochastically adding hidden nodes and links to the network, as shown in Fig. 1, keeping the mutation that yields the best performance.

Note that when using an add node mutation, an existing connection between two nodes is broken, giving place to a new node with two new connections. The link leading into this new node has unitary weight, and the other link getting out of this new node receives the same weight as the old connection.

Fig. 1 Topological mutations

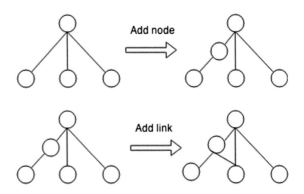

In this chapter, we will focus on NEAT and its applications on RL and BRL for a specific model-free control task, gait generation, instead of describing different evolutionary computation methods for function approximation on a general setup, as on [44]. Work therein develops XCS, a steady-state evolutionary method, based on learning classifier systems.

Several walking control studies have been performed on legged robots (bipeds [16, 17], quadrupeds [15, 30] and hexapods [6, 19]). Numerous six-legged robots have been developed in the last decades [5, 37]. Their main difficulty was to effectively deal with uneven terrains. Trying to cope with this problem, other works as [1, 2] have designed biologically inspired robots capable of walking on different kinds of surfaces. Closer to our application, [7] generates a free gait by applying reinforcement learning principles to the rectangular hexapod case, whose state space dimension is up to 46,656, due to the chosen representation.

Then, as QuadraTot [11] and Aracna [29], both open-source quadruped robots that have become popular among researchers, we also introduce Crabot, an hexapod robot with manipulation capabilities. The gait generation task is implemented in simulations of this robot by using FQI with NEAT as function approximator. Although this robot is already assembled, and the method proves to be faster than other approaches, the use of RL still represents a high cost on physical implementation, since gait generation methods based on reinforcement learning techniques enable the robot to learn over its own past failures, implying that any non-desirable action could lead to motor overheating or mechanical damage.

The ultimate objective of this six-legged robot, whose current version is shown on Fig. 2, is to provide a robotic platform capable of walking on uneven terrains and also manipulate objects at the same time. The next version of Crabot will be provided with grippers on the top of its legs.

The remainder of this chapter is as follows: Sect. 2 focuses on artificial neural networks that adjusts their topology and weights, in particular the method called NeuroEvolution of Augmenting Topologies (NEAT), which on its seminal paper was shown to outperform the best fixed-topology method on the pole balancing problem. Section 3 reviews some basic concepts on reinforcement learning, introducing some batch methods, e.g., experience replay and neural fitted Q iteration

Fig. 2 Crabot: the robotic crab

(NFQI), and studying how artificial neural networks can be used for function approximation on reinforcement learning methods, in order to tackle the curse of dimensionality. Section 4 shows how NEAT can be combined with Q-learning in the online approach, giving place to NEAT+Q, and combined with NFQI in the batch approach, obtaining NEFQI. Section 5 formulates the RL problem for gait generation and shows simulation results on an hexapod robot called Crabot. Finally Sect. 6 draws some final conclusions.

3 Topology and Weight Evolving Artificial Neural Networks

The main idea on neuroevolutive algorithms is to approximate a continuous non-linear function by applying evolutive algorithms on neural network models. These neuroevolutive algorithms optimize connection weights by using stochastic search, instead of gradient information, making them applicable to functions where gradient-based methods do not work well.

In principle, a fully connected network can approximate any continuous function without using any evolutive algorithm. The most important benefit from such approach is that network complexity does not affect speed and accuracy of learning when weights are being optimized by evolution instead of backpropagation.

The class of neuroevolutive algorithms that can evolve connection weights and network topology at the same time is known as a topology and weight evolving artificial neural network (TWEANN). In general, the initial population on a TWEANN is a collection of random topologies, which is problematic given that it could be possible that no path between a given input to an output is found.

A TWEANN can use either a direct or indirect encoding, where the prior scheme specifies in the genome every connection and node that will appear on the phenotype. The simplest implementation is based on a bit string representation that can be found in genetic algorithms, but there are several limitations, as having an impractical representation when the number of nodes becomes larger. On the other hand, indirect encoding specify rules for constructing this phenotype, like cellular encoding [13], whose genes can be reused several times during a phenotype (network) development. These genes requests cell divisions at different locations, just like in nature where a cell splits into more cells, in order to create an organism.

One of the main problems for neuroevolution is the permutations problem, which states that there is more than one way to represent a solution to a weight optimization problem with an artificial neural network. If we consider a TWEANN, the problem becomes worse, since such a network can represent similar solutions using totally different topologies.

In order to evolve minimal solutions, one of the most prominent TWEANN called neuroevolution of augmenting topologies (NEAT), starts with minimal topology, i.e., there are no hidden nodes and inputs are directly connected to outputs. Then, forthcoming structures are obtained by mutation, stochastically adding hidden nodes and links to the network, and keeping the mutation that yields the best performance. This approach is appealing given that in nature, complexity is developed over generations instead of being introduced at the beginning of evolution.

Note that when using an add node mutation, an existing connection between two nodes is broken, giving place to a new node with two new connections. The link leading into this new node has unitary weight, and the other link getting out of this new node receives the same weight as the old connection.

Recall that a genome corresponds to a linear representation of network connectivity, including a list of connection genes, specifying the appropriates input and output nodes, weight of connections and an innovation number for finding matching genes. Genes that do not match are either classified as disjoint (D) or excess (E), depending on whether they lie in the range of their parent's innovation numbers or not.

For the sake of clarity, Fig. 3 shows how a genome looks like, and Fig. 4 shows the corresponding phenotype. It can be seen, that the phenotype can be easily obtained from the genome, and conversely.

As stated on [38], especiation is introduced for preserving topological innovation. In order to help survive augmented structures that could be taken out due to

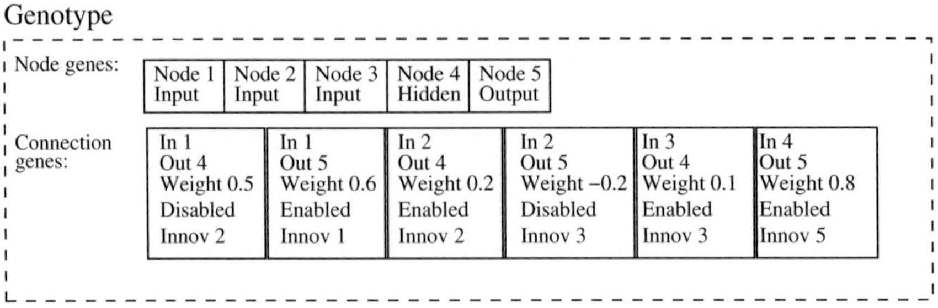

Genotype

Node genes:	Node 1 Input	Node 2 Input	Node 3 Input	Node 4 Hidden	Node 5 Output	
Connection genes:	In 1 Out 4 Weight 0.5 Disabled Innov 2	In 1 Out 5 Weight 0.6 Enabled Innov 1	In 2 Out 4 Weight 0.2 Enabled Innov 2	In 2 Out 5 Weight −0.2 Disabled Innov 3	In 3 Out 4 Weight 0.1 Enabled Innov 3	In 4 Out 5 Weight 0.8 Enabled Innov 5

Fig. 3 Genome structure

Fig. 4 Phenotype structure

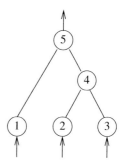

the initial loss of fitness when adding a new node, genomes are placed into species at each generation, depending on their compatibility. The compatibility distance, δ, of different structures is given by

$$\delta = c_1 \cdot \frac{E}{N} + c_2 \cdot \frac{D}{N} + c_3 \cdot \bar{W}, \tag{1}$$

where N stands for the number of genomes, and c_1, c_2 and c_3 are constants in order to adjust the importance of the number of excess, disjoint genes, and the average weight differences of matching genes respectively. This population division into species, based on topological similarity is possible by using historical markings.

NEAT keeps track of the historical origin of every gene, so if a new connection gene appears through structural mutation, a global innovation number is updated and assigned to that gene. The system knows exactly which genes match with other particular genes by using these innovation numbers.

Finally, to choose the best networks, the adjusted fitness f_i' for the i-th network is described by

$$f_i' = \frac{f_i}{\sum_j sh(\delta(i,j))}, \tag{2}$$

where f_i is the performance of the network, $\delta(i,j)$ is the distance from the i-th to every j-th network as on (1), and $sh()$ is the sharing function whose value depends on a threshold δ_t, such that

$$sh(\delta(i,j)) = \begin{cases} 0 & \text{if } \delta(i,j) > \delta_t, \\ 1 & \text{otherwise,} \end{cases} \tag{3}$$

Subsequently, species reproduce, i.e., new networks are created by removing members that perform worse over the population.

4   Reinforcement Learning

4.1   *Background*

Reinforcement learning (RL) tackles the problem of an agent that learns while interacting with the environment. It has to decide which action a to execute on its current state s, which transfers the agent to another state s' receiving a reward (reinforcement signal) r that provides a quantification of how desirable was that choice.

In formal terms, a RL problem can be formulated as a MDP [39], composed by a tuple (S, A, T, R) where

- S: denotes the set of all possible states.
- A: is set of all the actions the agent can execute.
- T: $S \times A \times S \rightarrow [0, 1]$ is a state transition function, which gives the probability that when the agent is in state s and executes action a, the agent will be transferred to another state s'.
- R: $S \times A \rightarrow \mathbb{R}$ is the scalar (real-valued) reward function.
- π: $S \rightarrow A$ denotes the mapping from states to action, describing the policy the agent should take given a certain state.

As indicated earlier, the task of the agent is to learn the sequence of actions (therefore the optimal policy, π^*) that leads to maximize the expected sum of all the rewards received in the long-term. That is, the agent maximizes the return, R_t.

Let R_t be the return, defined as the discounted sum of rewards that the agent will obtain from time t, given by

$$R_t = \sum_{k=0}^{n-1} \gamma^k r_{t+1+k}, \tag{4}$$

where γ stands for the discount factor, with $0 \leq \gamma < 1$, and r_{t+1} stands for the (scalar) reward obtained for executing action a_t in state s_t. By re-arranging terms on (4), a recursive expression is obtained as

$$= r_{t+1} + \sum_{k=0}^{n-1} \gamma^{k+1} r_{t+2+k} \tag{5}$$

$$= r_{t+1} + \gamma R_{t+1}$$

Two quantifications for the expected return are defined, the value function and action value function, V^π and Q^π respectively. Value function is defined as the expected return when the agent is on state s_t at time t,

$$V^{\pi}(s) = E_{\pi}\{R_t|s_t = s\}, \tag{6}$$

while the action value function is defined as the expected return when the agent executes a_t on state s_t at time t following policy π,

$$Q^{\pi}(s,a) = E_{\pi}\{R_t|s_t = s, a_t = a\}, \tag{7}$$

where both functions are clearly related, as

$$V^{\pi}(s) = E_{a|s}\{Q^{\pi}(s,a)\}. \tag{8}$$

The policy π that maximizes the expected discounted return from each state is called an optimal policy, and is denoted by π^*. The value function corresponding to an optimal policy is called the optimal value, and is denoted by V^*. While there may exist multiple optimal policies, the optimal value is unique [3] and may be computed by solving the Bellman optimality equation

$$V^*(s) = \max_{a \in A} \sum_{s'} Pr\{(s,a,s')\}(R(s,a) + \gamma V^*(s')), \tag{9}$$

where $Pr\{(s,a,s')\}$ stands for the probability of being transferred to some state s' when executing action a on state s. One of the two most popular and convergent dynamic programming algorithms, is the value iteration (VI) algorithm, whose pseudocode is shown in Algorithm 1, and works by arbitrarily initializing a value estimate, for all $s \in S$.

Algorithm 1 Value Iteration algorithm

1: $i = 0$
2: $\hat{V}_0(s) = 0 \quad \forall s \in S$
3: **while** $\Delta > \varepsilon$ **do**   ▷ (for a small positive number ε)
4:     **for** each $s \in S$ **do**
5:        $\hat{V}_{i+1}(s) = \max_{a \in A} Pr\{(s,a,s')\}(R(s,a) + \gamma V(s'))$
6:     **end for**
7:     $\Delta = \|\hat{V}_{i+1} - \hat{V}_i\|$
8:     $i = i + 1$
9: **end while**
10: **return** policy π ▷ such that maximizes (9)

This algorithm involves the computation of some norm of the difference between two subsequent value estimates. Several theoretical results exist, which provide bounds for the performance of π^* as a function of ε for different norms, including the maximum norm $\| \cdot \|_{\infty}$ [45] and weighted l_2 norms [32].

Another algorithm, Policy Iteration (PI), shown in Algorithm 2, works by the process of policy improvement. Specifically, it starts with some initial random policy. Then, at each successive iteration, it evaluates the value function for the current policy, and performs a policy improvement step in which a new policy is generated by selecting a greedy action at each state, with respect to the values of the current policy. If the improved policy is the same as the policy improved upon, then we are assured that the optimal policy has been found.

Most successful reinforcement learning algorithms are descended from one of the two dynamic programming algorithms described above, VI and PI. However, there are two major features distinguishing the RL setting from the traditional decision theoretic setting.

First, while in decision theory it is assumed that the environment model is fully known, in RL no such assumption is made. Second, the learning process is usually assumed to take place online, namely, concurrently with the accumulation of actual (or simulated) data acquired by the learning agent as it explores its environment. These two features make RL a significantly more difficult challenge, and place serious constraints on any potential RL algorithm. Probably the two best known RL algorithms, TD(λ) [39] and Q-learning [42], serve well to demonstrate how RL methods handle these constraints. For simplicity, we assume that the state and action spaces are finite, and that the state values, or state-action values are stored explicitly in a lookup table.

Algorithm 2 Policy Iteration algorithm

1: $\hat{\pi}_0(s) =$ some random action $\forall a \in A$
2: $\hat{V}_0(s) = 0$ $\forall s \in S$ ▷ arbitrarily
3: $i = 0$
4: **while** $\Delta > \varepsilon$ **do**   ▷ (for a small positive number ε)
5:      **for** each $s \in S$ **do**
6:          $\hat{V}_{i+1}(s) = \max_{a \in A} Pr\{(s,a,s')\} (R(s,a) + \gamma V(s'))$
7:      **end for**
8:      $\Delta = \|\hat{V}_{i+1} - \hat{V}_i\|$
9:      $i = i+1$
10: **end while**
11: *policy_stable* $= true$
12: $i = 0$
13: **for** each $s \in S$ **do**
14:     $\hat{\pi}_{i+1}(s) = argmax_{a \in A} Pr\{(s,a,s')\} (R(s,a) + \gamma V(s'))$
15:     **if** $\hat{\pi}_i(s) \neq \hat{\pi}_{i+1}(s)$ **then**
16:         *policy_stable* $= false$
17:     **end if**
18: **end for**
19: **if** *policy_stable* **then**
20:     **return** policy $\hat{\pi}_i$
21: **else**
22:     go to line 3
23: **end if**

TD(λ) is aimed at evaluating the value function for a given policy π. The input to the algorithm is a sequence of state-rewards couples, generated by the MDP controlled by the policy π. The idea in TD(λ) is to gradually improve value estimates by moving them towards the weighted average of multi-step lookahead estimates, which take into account the observed rewards. In the simplest case, of $\lambda = 0$, this amounts to moving the value estimate of the current state, $\hat{V}(s)$, toward the one-step lookahead estimate $R(s, a) + \gamma \hat{V}(s')$. Temporal difference (TD) methods avoid making direct use of the transition model by sampling from it.

The pseudocode for TD(0) is given in Algorithm 3, where the update term $R(s, a) + \gamma \hat{V}(s') - \hat{V}(s)$ is known as the temporal difference.

Algorithm 3 Tabular TD(0) algorithm

1: $\hat{V}(s) = 0$ $\forall s \in S$
2: **for** each episode **do**
3:     **Observe** $s, R(s, a), s'$
4:     $\hat{V}(s) = \hat{V}(s) + \alpha \left(R(s, a) + \gamma \hat{V}(s') - \hat{V}(s) \right)$
5: **end for**
6: **return** $\hat{V}$

In many cases, RL tasks are naturally divided into learning episodes. In such episodic learning tasks, the agent is placed at some (typically randomly chosen) initial state, and is then allowed to follow its policy until it reaches a terminal state, where the episode terminates and a new one may begin. A terminal state is modeled as a state with zero reward and with only self transitions, for any action. If s' is terminal in the TD(0) update, as well as in algorithms presented in the sequel, we define $\hat{V}(s') = 0$.

As any value estimation algorithm requires a policy improvement step, and it involves knowledge of the transition model which is typically unavailable. This is often solved by introducing state-action values (also known as Q-values), rather than just state values.

For a given policy π, the Q-value for the state-action pair (s, a) is the expected discounted return over all trajectories starting from s, for which the first action is a, and with all subsequent actions chosen according to π. The optimal state value V^* is related with the optimal state-action value Q^* as

$$V^*(s) = \max_{a \in A} Q^*(s, a). \tag{10}$$

As on TD algorithms, where expectations were replaced with actual samples, we can by analogy derive the Q-learning algorithm, which can be viewed as an asynchronous, stochastic version of VI. The pseudocode is shown on Algorithm 4.

Algorithm 4 Q-learning algorithm

1: Initializa arbitrarily $\hat{Q}(s,a)$ $\forall s \in$ S, $a \in$ A
2: **for** each episode **do**
3:     **Observe** $s, a, R(s,a), s'$
4:     $\hat{Q}(s,a) = \hat{Q}(s,a) + \alpha \left(R(s,a) + \gamma \max_{a'} \hat{Q}(s',a') - \hat{Q}(s,a) \right)$
5: **end for**
6: **return** $\hat{Q}$

Thus, by assuming $\hat{Q} = Q^*$ (i.e., Q-learning has converged to the optimum), then an optimal action for each state can be easily computed by a single maximization operation

$$\pi^*(s) = arg \max_a Q^*(s,a), \tag{11}$$

with arbitrarily broken ties.

TD and Q-learning algorithms, as well as many other algorithms relying on a look-up table representation, are useful in providing a proof-of-concept. However, real world problems can rarely be solved using such representations, due to the large, and sometimes infinite state and action spaces which characterize such problems. Since a tabular representation is unfeasible, it is necessary, in such problems, to use some form of function approximation to represent the value function, which will be introduced in a subsequent section.

4.2 Batch Reinforcement Learning

When we talk about batch reinforcement learning (BRL), we refer to one of the current lines of research in the field of reinforcement learning (RL), also concerned about solving sequential decision problems modelled by a markovian decision process (MDP). Given the nature of these problems, as the intuition may suggest, the scope of this type of learning has extended to areas like robotics [22].

As the reader may infer, the need for function approximation naturally arises when the state space becomes larger. This is the same reason that creates the problem that BRL tackles, given that these online algorithms require that all state-action pairs to be continuously updated for ensuring convergence.

BRL approach aims to collect a bunch of experiences and then to use them for updating action influences, instead of updating the action value function in an incremental way. In this batch framework, algorithms like experience replay (ER) or fitted Q iteration (FQI) [8] can be found, where the latter has been taken as basis for derivations like neural FQI [36] or FQI with alternative update rules like Q-Batch [4].

In order to understand the difference between the incremental update from online algorithms and simultaneous update of batch algorithms, consider two consecutive transitions (s, a, r, s'), (s', a, r, s'') and the classical Q(0) online algorithm. When $Q(s', a)$ is computed using update rule on Algorithm 4, this change won't be backpropagated to $Q(s, a)$ nor any of the state-action pairs preceding s', being updated just when those states are visited again.

In the pure batch reinforcement learning approach, the agent doesn't interact with the environment while the learning phase is taking place. Yet, in growing batch reinforcement learning, which most of the modern batch algorithms are based on, the task of collecting transitions and learning from these are alternated for improving the exploration policy.

Algorithm 5 Batch reinforcement learning procedure

1: Initialize $Q(s, a)$ arbitrarily $\forall\, s \in$ S, $a \in$ A
2: Initialize batch of experiences D as an empty set
3: **repeat**
4:      **repeat**[for each episode]
5:          **repeat**[for each step t of the current episode]
6:              Identify current state s_t
7:              Choose a suitable action a_t in state s_t using policy derived from Q
8:              Observe r_{t+1} and s_{t+1} when taking action a_t
9:              Add experience $(s_t, a_t, r_{t+1}, s_{t+1})$ on batch D
10:             $s_t \leftarrow s_{t+1}$
11:         **until** some stop criteria is met
12:     **until** all episodes are visited
13:     Update Q values
14:     Forget m experiences from batch D
15: **until** action value function convergence is reached

Algorithm 5 describes the procedural form of a (growing) BRL approach independently of the algorithm used for updating Q-values, as shown in [18]. Note that when $m = |D|$ experiences are forgotten, growing BRL is reduced to pure BRL.

Moreover, [18] states that it is better (for their task) to use all the experiences gathered so far. This means that if every batch consists on experiences from 20 episodes, then the first updates of Q estimations will consists on experiences from those 20 episodes. The second time these updates are computed it will consists on experiences from 40 episodes (the first 20 and the newest 20), and so on, which is an extremely memory consuming task.

One of the basic BRL algorithms, first introduced by [27], experience replay (ER) aim to improve the speed of convergence of the action value function by replaying observed transitions repeatedly just as if they were new observations. Algorithm 6 shows its procedural form.

Algorithm 6 Experience Replay procedure

1: **repeat**[for each training iteration]
2:　　**repeat**[for each transition $(s_i, a_i, r_{i+1}, s_{i+1})$ on D]
3:　　　　Update $Q(s_i, a_i)$ by using

$$Q(s_i, a_i) \leftarrow (1-\alpha) Q(s_i, a_i) + \alpha \left(r_{i+1} + \gamma \max_a Q(s_{i+1}, a) \right)$$

4:　　**until** all transitions on D are visited
5: **until** all training iterations are executed

Note that, this algorithm does computes several times the updates of Q-learning on collected transitions as an offline algorithm would do, thus speeding up the propagation of Q values to preceding states. Then, the system is allowed to collect new transitions for improving those previously computed estimates.

Neural fitted Q iteration (NFQI) is one of the most popular BRL algorithms, introduced in [8], which consists on iterations between two steps:

- Update the empty set P^{k+1} appending tuples of the form $(s, a, \bar{q}^{k+1}(s, a))$ for every experience on (s, a, r, s'), where

$$\bar{q}^{k+1}(s, a) = r + \gamma \max_{a' \in A} \hat{Q}^k(s', a') \tag{12}$$

- Compute the approximation $\hat{Q}^{k+1}$, based on the training set P^{k+1},

where k represents the k-th iteration of FQI algorithm, and γ is the discount factor as defined earlier in Sect. 3.

Work in [8] uses randomized trees for approximating Q^{k+1} at every iteration, while another popular supervised algorithm refers to artificial neural networks, as they are known to be capable of representing a wide variety of functions given the appropriate parameters. This version is known as neural fitted Q iteration, as seen on [36], which makes use of resilient backpropagation, although other methods could be applied.

The procedural form of NFQI is shown in Algorithm 7.

Algorithm 7 (Neural) Fitted Q Iteration procedure

1: Reset training iterations counter $k = 0$
2: **for** each training iteration **do**
3:　　Create an empty set P^{k+1}
4:　　**for** each tuple $(s_i, a_i, r_{i+1}, s_{i+1})$ on D **do**
5:　　　　Update $\bar{q}^{k+1}(s_i, a_i)$ by (12)
6:　　　　Append $(s_i, a_i, \bar{q}^{k+1})$ to P^{k+1}
7:　　**end for**
8:　　Adjust weights of ANN to approximate Q^{k+1} based on P^{k+1}
9:　　$k = k+1$
10: **end for**

The main idea behind introducing artificial neural networks on this setup refers to the fact, as stated earlier, that each state-action pair has to be visited enough times for a Q-function to converge, which is impractical when the state (or action) space is high-dimensional (intractable for continuos spaces). Consequently, a function approximator is needed, in order to approximate the action-value function, Q(s, a), or the value function V(s). There are non-linear function approximators, or parametric approximations, such as a linear combination of basis functions

$$\hat{V}(s) = \sum_{i=1}^{k} w_i \phi_i(s)$$
$$= w^{\mathsf{T}} \Phi(s),$$

(13)

where $\Phi(s) = \{\phi_1(s), \ldots, \phi_k(s)\}$ is a set of basis functions, with $\phi_i(s)$ defined as the value of feature i in state s, while $w = \{w_1, \ldots, w_k\}$ is a set of scalar weights.

There are works focused on estimating the weights for a given set of basis functions, as in [34], or finding suitable basis functions as in [23]. On the other hand, non-linear approaches found in literature for function approximation methods, refer to fuzzy logic [12], artificial neural networks [28] or decision trees [35].

5   RL with Evolutionary Function Approximation

When reinforcement learning problems are being tackled by evolutionary methods, they usually evolve a population of action selectors, in such a way that each of these action selectors remains fixed on the fitness evaluation step. The main objective of evolutionary function approximation is to use temporal difference methods to update value functions on the fitness evaluation step, evolving function approximators that are better able to learn by temporal difference instead of evolving actions selectors directly [43].

Also, synergistic effects between evolution and learning can be achieved by applying evolutionary function approximation, and how this is accomplished depends on whether a Lamarckian or Darwinian approach is being applied. The prior implies that changes made during a generation are backed up into the original genotype, in order to be used on the next generation. A Darwinian implementation discards these changes and always takes the original genotype, just like in the first generation.

Regardless the nature of biological systems reflecting a Lamarckian or a Darwinian approach, it can be seen that the benefits of the prior are to prevent each generation from repeating the same learning. The latter allow us to avoid relying on continued altered versions of the genome, that may not grow as expected.

Note that all results presented on this chapter make use of a Darwinian implementation.

5.1   Neat+Q

As indicated earlier, NEAT typically optimize action selectors. Reinterpretation of its output values is needed in order to optimize value functions instead. Weights evolves are updated on the fitness evaluation, since input and output layers of NEAT networks are already the same as networks normally used as function approximators on Q-learning, which would be one input for each state feature and one output for each action.

Algorithm 8 NEAT+Q procedure

1: Create new population with random networks
2: **for** each generation **do**
3:    **for** each network in population **do**
4:       **for** each episode **do**
5:          Observe $s, a, R(s,a), s'$
6:          Update fitness of current network $f = f + R(s,a)$
7:          Adjust weights of current network to approximate toward targets $(R(s,a) + \gamma \max_{a'} Q(s',a'))$
8:       **end for**
9:    **end for**
10:   $p = 0$
11:   **while** $p < population\_size$ **do**
12:      Make a new network from fit parents
13:      Add a (random) node with probability p_n
14:      Add a (random) link with probability p_l
15:      $p+ = 1$
16:   **end while**
17: **end for**

The benefit of using NEAT instead of fixed-topology neural networks is straightforward, since NEAT will learn an effective representation by exploring the space of networks that leads to better performance, instead of relying on a hand-tuned topology network expecting to perform well. The procedural form of the mix between online (classical) Q-learning and NEAT, is shown on Algorithm 8, as can be found on [43].

Note that, in order to choose a suitable action given the state and approximation of Q obtained by the current network, several mechanisms could be introduced for tackling the exploration-exploitation dilemma. The most common approach uses an ε-greedy technique, which consists on choosing a random (or exploratory) action with probability ε, and the action that yields a maximum action-value function, with probability $1 - \varepsilon$ [39].

5.2   NEFQI

NEAT is typically used as a neural network action selector, i.e., it optimizes directly action selections by letting the input nodes depict the state features, and the output nodes depict the possible actions. In this work, NEAT is used as a function approximator for the action-value function, so the network is fed by the state-action pair and its output corresponds to the approximated action-value function, $\hat{Q}$.

The procedural form of neuro-evolutive fitted Q iteration (NEFQI) is shown in Algorithm 9.

When using NEFQI within the BRL setup, the network with best performance (fitness) resulting from all the training phase, is used for evaluating the action-value function in order to choose a suitable action. The fact of repeatedly visiting the transitions samples of the batch for every network on the population at each generation, is the essence of BRL, since its objective is to use collected data efficiently.

Note that fitness for the i-th network is given by

$$f_i = \sum_{(s,a,r,s') \in D} r + \gamma \max_{a'} \hat{Q}_i(s', a'), \tag{14}$$

where $\hat{Q}_i$ corresponds to the approximated action-value function by the current network, when its weights have converged to some appropriate values. Note that the intuition may suggest that this expression is the same for every network, since r is given by the interactions of the agent with its environment, but the last part of (14) depends on the approximation obtained by each network, and corresponds to the term to be maximized, since it is desirable to maximize the discounted long-term expected reward.

Algorithm 9 Neuro-Evolutive Fitted Q Iteration procedure

1: Create new population with random networks
2: **for** each generation **do**
3:    **for** each network in population **do**
4:        Reset training iterations counter $k = 0$
5:        **for** each training iteration **do**
6:            Create an empty set P^{k+1}
7:            **for** each tuple $(s_i, a_i, r_{i+1}, s_{i+1})$ on D **do**
8:                Update $\bar{q}^{k+1}(s_i, a_i)$ by (12)
9:                Append $(s_i, a_i, \bar{q}^{k+1})$ to P^{k+1}
10:               **if** final training iteration **then**
11:                   Update fitness of current network $f = f + \bar{q}^{k+1}$
12:               **end if**
13:           **end for**
14:           Adjust weights of current network to approximate Q^{k+1} based on P^{k+1}
15:           $k+ = 1$
16:       **end for**
17:   **end for**
18:   $p = 0$
19:   **while** $p < population\_size$ **do**
20:       Make a new network from fit parents
21:       Add a (random) node with probability p_n
22:       Add a (random) link with probability p_l
23:       $p+ = 1$
24:   **end while**
25: **end for**

The aim of NEFQI is to make an efficient use of collected data, regardless the chosen topology for the neural network, since it would start with no hidden layers and then its structure will evolve accordingly, noting that the fitness of the current network will be evaluated just the last time the whole batch is visited, in order to ensure that the different networks are compared once all the weights have been adjusted after all the action-value samples have been propagated.

6 Experimental Results

6.1 The Robotic Crab

Crabot has been constantly growing in terms of design. It has been designed and built at Centro de Robótica (Robotics Centre) at Universidad Técnica Federico Santa Maria, Chile, an undergraduate and graduate students group focused on robotics research and development. Its current version is shown on Fig. 5. The main difference with the first version is the improved design on its legs, giving priority to its locomotion capabilities over the vision system that it used to have. The ultimate objective of this six-legged robot is to provide a robotic platform capable of walking on uneven terrains and manipulate objects at the same time. Consequently, the next version of Crabot will be provided with grippers on the top of its legs.

Each degree of freedom, is implemented by a Dynamixel smart servo, with maximum torque of 1.8 Nm. Each servo has an integrated controller, and automatic information provided, such as internal temperature, current consumption and torque measures.

Although the final version of Crabot is planned to have five degrees of freedom, both the current and the first version have four degrees of freedom per leg. Nevertheless, just three of them are used for locomotion. In this experiment, the fourth motor is used as a rotatory shoulder for manipulation tasks, and it is fixed into the main hexagonal body, as can be seen on Fig. 6. The fifth motor, yet to be added, will be used for the grip.

Fig. 5 Current version of Crabot

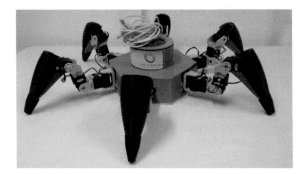

Fig. 6 Degrees of freedom of
the first two prototypes

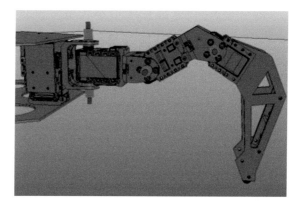

The benefits of adding grippers into the final version are obvious, since the robot would be able to manipulate objects with its legs, just as a crab would do. Unlike the biological inspiration of this robot, Crabot would be able to manipulate objects with any of its legs, as long as it can perform a dynamical stable walking pattern if needed. This also represents a big challenge, because in addition to the high complexity, given the number of legs and degrees of freedom per leg, it requires that the robot can adapt its walking pattern regardless of the method or type of learning that is being used, learning that legs can be used both for walking or manipulation depending on the circumstances. Upon completion, the final version is planned to be released with further details about its design, so the community can build their own with a 3D printer. The robot can be used as a benchmark robot for future research.

6.2 Simulated Environment

Figure 7 shows the first prototype of Crabot on the educational version of V-REP [9], a 3D simulated environment that allows a fast and customizable dynamics calculation within a user-friendly programming framework.

Fig. 7 Simulated model of
Crabot on V-REP

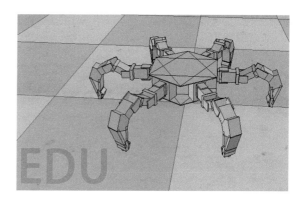

Crabot is expected to learn to walk on a plain simple surface after enough iterations over the learning process. The robot always starts with its center of gravity at $(0,0)$ on cartesian coordinates, and stops interacting with the environment, i.e., stops collecting transition samples, when any part of the robot except its legs collides with the surface, or 4 s are elapsed, whatever happens first. According to Algorithm 5, $m = |D|$, so all samples from the batch are removed once a neural network has been chosen for NFQI or NEFQI.

The state is given by the position of every of the eighteen locomotion motors,

$$s = (m_1, m_2, \ldots, m_{18}), \tag{15}$$

where m_i is the i-th motor, noting that the two motors that are closest to the main body for each leg, are constrained to be in the range between $[-45°, 45°]$, while the others are limited to the range $[-90°, 90°]$.

Although we work under a continuous state space representation, the action space has been discretized, such that at each step, the i-th motor can move clockwise, counter-clockwise or stay in its current position. Even if we have a good approximation of the state-action value function, it is not trivial to find the action that yields a highest value when dealing with a continuous action space.

Since there are different limits for every motor, the rotations that each motor can perform at each step are different too, so the two motors that are closest to the main body for each leg, can move 4.5° at each step, while the others can move 9° at each step.

The reward at time t is set to

$$r_t = \begin{cases} -1 & \text{robot collide,} \\ dist(s_t, s_{t-1}) & \text{otherwise,} \end{cases} \tag{16}$$

where $dist(s_t, s_{t-1})$ stands for the distance travelled by the center of gravity of the robot when moving from state s_{t-1} to state s_t.

The discount factor was arbitrarily set to $\gamma = 0.8$. Recall from Sect. 3, that a higher discount factor implies that a higher number of future reinforcement signals will be considered for evaluating performance. Then, NFQI and NEFQI were compared with 20 training iterations per batch, collecting samples into the batch after every episode, as shown in Fig. 8, with NFQI using a random neural network in terms of its (fixed) topology, and using the same method as NEFQI for adjusting its weights.

For NEFQI, coefficients on (1) are set to $c_1 = 1$, $c_2 = 1$ and $c_3 = 0.4$, which implies that the number of excess and disjoint genes are equally more important than the average weight differences of matching genes for computing compatibility distance between networks. The distance threshold used on (3) is set to $\delta_t = 6$, and the population size is up to 50.

The simulation, according to Algorithm 9 was executed over 50 generations, with equal probabilities of adding a new node or a new link between existing nodes, $p_n = p_l = 0.08$.

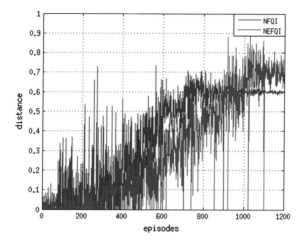

Fig. 8 Distances travelled by Crabot while learning

Figure 8 shows that NEFQI has better performance than NFQI when convergence is reached, but NEFQI convergence rate is slower because of the topological modifications on the network. Although the reward corresponds to the euclidean distance travelled by Crabot, note that distances cannot be negative, so episodes when the distance is zero contains episodes when the robot collided with the surface, and also when the robot tried useless actions that were not enough to make it move from the starting point within the time window of 4 s.

Then, Figs. 9 and 10 shows the evolution of the robot position at the learning episode 160 and episode 1200 respectively. It can be seen that in the first episodes,

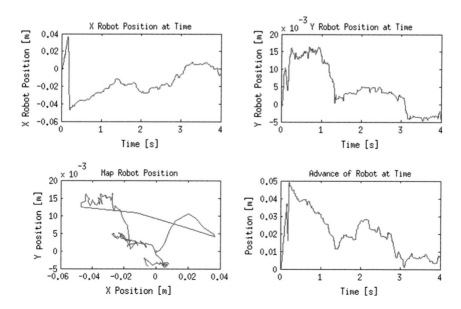

Fig. 9 Robot position at episode 160

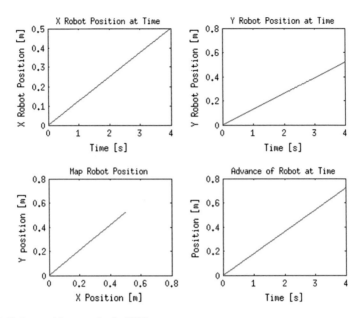

Fig. 10 Robot position at episode 1200

as on episode 160, it is usual that the robot cannot move around because of collisions, but finally when convergence is met, the robot learns to walk on a straight path, but yet it is not clear whether a faster gait could be generated or not. Note that no preference was given for a predefined path, if that would be the case, the reward function would have to be modified accordingly.

As a concluding remark, it can be seen that using NEAT on NFQI as a function approximator allow the hexapod to achieve better performance than a fixed topology ANN, at a slower convergence rate. This is an important aspect for future work, since short changes on the environment such as small portions of uneven surface could be considered as disturbances on the learning process. On the other hand, if the whole surface is suddenly changed, the hexapod behavior is expected to adapt accordingly.

7 Conclusions

This chapter reviewed the method known as neuro-evolution of augmenting topologies which corresponds to a particular class of topology and weight evolving ANN, finding a suitable network for a given problem starting from a simple artificial neural network with no hidden layer. We used NEAT as a function approximator for NFQI, a batch reinforcement learning method that typically uses a fixed-topology ANN for dealing with a continuous state space representation.

Although allowing topology of neural networks to evolve, proved to have slower convergence than just adjusting its weights, NEFQI reached better performance than NFQI at last, showing that as suggested, function approximation is a crucial step in RL techniques.

This chapter also presented Crabot, an hexapod robot biologically inspired by a crab. We tested NEFQI and NFQI on a simulated version of this robot for a gait generation task, learning to walk on a simple plain surface.

Given that these gait generation methods require to fail in order to avoid such states in future trials, it is still too dangerous to implement on the physical robot, since it could damage mechanical pieces or motors. This could be avoided by hand-tuning constraints on the set of possible actions, but there is still a possibility of getting unforeseen behavior or we may end up forbidding a combination of motor movements that could eventually lead to a better performance.

Testing and analysis of improvement of adding NEAT into other online and batch RL algorithms remains as future work, as well as the documentation of the final version of Crabot upon completion of its ultimate design, with grippers on top of its legs, in order to introduce this robot as a new robotic benchmark problem for manipulation and locomotion tasks on multi-legged robots.

References

1. Altendorfer, R., Moore, N., Komsuoglu, H., Buehler, M., Brown Jr, H., McMordie, D., Saranli, U., Full, R., Koditschek, D.E.: Rhex: a biologically inspired hexapod runner. Auton. Robots **11**(3), 207–213 (2001)
2. Beer, R.D., Quinn, R.D., Chiel, H.J., Ritzmann, R.E.: Biologically inspired approaches to robotics: what can we learn from insects? Commun. ACM **40**(3), 30–38 (1997)
3. Bertsekas, D.P., Bertsekas, D.P.: Dynamic programming and optimal control, vol. 1. Athena Scientific, Belmont (1995)
4. Cunha, J., Lau, N., Neves, A.J.R.: Q-batch: initial results with a novel update rule for batch reinforcement learning. In: Advances in Artificial Intelligence-Local Proceedings, XVI Portuguese Conference on Artificial Intelligence. Azores pp. 240–251 (2013)
5. Devjanin, E.A., Gurfinkel, V.S., Gurfinkel, E.V., Kartashev, V.A., Lensky, A.V., Yu Shneider, A., Shtilman, L.G.: The six-legged walking robot capable of terrain adaptation. Mech. Mach. Theor. **18**(4), 257–260 (1983)
6. Duan, X., Chen, W., Yu, S., Liu, J.: Tripod gaits planning and kinematics analysis of a hexapod robot. In: Control and Automation, 2009. ICCA 2009. IEEE International Conference on, pp. 1850–1855, IEEE (2009)
7. Erden, M.S., Leblebicioğlu, K.: Free gait generation with reinforcement learning for a six-legged robot. Robot. Auton. Syst. **56**(3), 199–212 (2008)
8. Ernst, D., Geurts, P., Wehenkel, L.: Tree-based batch mode reinforcement learning. J.Mach. Learn. Res., 503–556 (2005)
9. Freese, M., Singh, S., Ozaki, F., Matsuhira, N.: Virtual robot experimentation platform v-rep: a versatile 3d robot simulator. Simulation, modeling, and programming for autonomous robots, pp. 51–62. Springer, Berlin (2010)
10. Ghanbari, A., Vaghei, Y., Noorani, S., Reza, S.M.: Reinforcement learning in neural networks: a survey. Int. J. Adv. Biol. Biomed. Res. **2**(5), 1398–1416 (2014)

11. Glette, K., Klaus, G., Zagal, J.C., Torresen, J.: Evolution of locomotion in a simulated quadruped robot and transferral to reality. In: Proceedings of the Seventeenth International Symposium on Artificial Life and Robotics (2012)
12. Glorennec, P.Y., Jouffe, L.: Fuzzy Q-learning. In: Fuzzy Systems, 1997., Proceedings of the Sixth IEEE International Conference on, vol. 2. pp. 659–662, IEEE (1997)
13. Gruau, F.: Genetic synthesis of modular neural networks. In: Proceedings of the 5th International Conference on Genetic Algorithms, pp. 318–325. Morgan Kaufmann Publishers Inc. (1993)
14. He, P., Jagannathan, S.: Reinforcement learning-based output feedback control of nonlinear systems with input constraints. IEEE Trans. Syst. Man Cybern. B Cybern. 35(1), 150–154 (2005)
15. Hirose, S., Fukuda, Y., Yoneda, K., Nagakubo, A., Tsukagoshi, H., Arikawa, K., Endo, G., Doi, T., Hodoshima, R.: Quadruped walking robots at tokyo institute of technology: design, analysis, and gait control methods. IEEE Robot. Autom. Mag. 16(2), 104–114 (2009)
16. Huang, Q., Yokoi, K., Kajita, S., Kaneko, K., Arai, H., Koyachi, N., Tanie, K.: Planning walking patterns for a biped robot. IEEE Trans. Robot. Autom. 17(3), 280–289 (2001)
17. Kajita, S., Morisawa, M., Miura, K., Nakaoka, S., Harada, K., Kaneko, K., Kanehiro, F., Yokoi, K.: Biped walking stabilization based on linear inverted pendulum tracking. In: Intelligent Robots and Systems (IROS), 2010 IEEE/RSJ International Conference on, pp. 4489–4496. IEEE (2010)
18. Kalyanakrishnan, S., Stone, P.: Batch reinforcement learning in a complex domain. In: Proceedings of the 6th International Joint Conference on Autonomous Agents and Multiagent Systems, p.94. ACM (2007)
19. Kamikawa, K., Arai, T., Inoue, K., Mae, Y.: Omni-directional gait of multi-legged rescue robot. In: Robotics and Automation, 2004. Proceedings. ICRA'04. 2004 IEEE International Conference on, vol. 3, pp. 2171–2176. IEEE (2004)
20. Kiumarsi, B., Lewis, F.L., Modares, H., Karimpour, A., Naghibi-Sistani, M.B.: Reinforcement Q-learning for optimal tracking control of linear discrete-time systems with unknown dynamics. Automatica 50(4), 1167–1175 (2014)
21. Kiumarsi-Khomartash, B., Lewis, F.L., Naghibi-Sistani, M.B., Karimpour, A.: Optimal tracking control for linear discrete-time systems using reinforcement learning. In: Decision and Control (CDC), 2013 IEEE 52nd Annual Conference on, pp. 3845–3850. IEEE (2013)
22. Kober, J., Bagnell, J.A., Peters, J.: Reinforcement learning in robotics: a survey. Int. J. Robot. Res. 32(11), 1238–1274 (2013)
23. Konidaris, G., Osentoski, S., Thomas, P.S.: Value function approximation in reinforcement learning using the fourier basis. In: AAAI (2011)
24. Kosslyn, S.M., Kosslyn, S.: Top brain, bottom brain: surprising insights into how you think. Simon and Schuster, New York (2013)
25. Lange, S., Gabel, T., Riedmiller, M.: Batch reinforcement learning. In: Reinforcement Learning, pp. 45–73. Springer, Berlin (2012)
26. Lewis, F.L., Liu, D.: Reinforcement learning and approximate dynamic programming for feedback control, vol. 17. Wiley, New York (2013)
27. Lin, L.J.: Self-improving reactive agents based on reinforcement learning, planning and teaching. Mach. Learn. 8(3–4), 293–321 (1992)
28. Lin, L.J.: Reinforcement learning for robots using neural networks. Technical report, DTIC Document (1993)
29. Lohmann, S., Yosinski, J., Gold, E., Clune, J., Blum, J., Lipson, H.: Aracna: an open-source quadruped platform for evolutionary robotics. Artif. Life 13, 387–392 (2012)
30. Ma, S., Tomiyama, T., Wada, H.: Omnidirectional static walking of a quadruped robot. IEEE Trans. Robot. 21(2), 152–161 (2005)
31. Modares, H., Lewis, F.L.: Online solution to the linear quadratic tracking problem of continuous-time systems using reinforcement learning. In: Decision and Control (CDC), 2013 IEEE 52nd Annual Conference on, pp. 3851–3856. IEEE (2013)

32. Munos, R.: Error bounds for approximate policy iteration. In: Proceedings of the 20th International Conference on Machine Learning (ICML-03), pp. 560–567 (2003)
33. Nakamura, Y., Mori, T., Sato, M., Ishii, S.: Reinforcement learning for a biped robot based on a cpg-actor-critic method. Neural Netw. **20**(6), 723–735 (2007)
34. Parr, R., Li, L., Taylor, G., Painter-Wakefield, C., Littman, M.L.: An analysis of linear models, linear value-function approximation, and feature selection for reinforcement learning. In: Proceedings of the 25th International Conference on Machine Learning, pp. 752–759. ACM (2008)
35. Pyeatt, L.D., Howe, A.E., et al.: Decision tree function approximation in reinforcement learning. In: Proceedings of the Third International Symposium on Adaptive Systems: Evolutionary Computation and Probabilistic Graphical Models, vol. 2. pp. 70–77 (2001)
36. Riedmiller, M.: Neural fitted Q iteration–first experiences with a data efficient neural reinforcement learning method. In: Machine Learning: ECML 2005, pp. 317–328. Springer, Berlin (2005)
37. Schmucker,U., Schneider, A., Ihme, T.: Hexagonal walking vehicle with force sensing capability. In: Proceedings of 6th International Symposium on Measurement and Control in Robotics. Brussel, pp. 354–359 (1996)
38. Stanley, K.O., Miikkulainen, R.: Evolving neural networks through augmenting topologies. Evol. Comput. **10**(2), 99–127 (2002)
39. Sutton, R.S., Barto, A.G.: Introduction to reinforcement learning. MIT Press, Cambridge (1998)
40. Sutton, R.S., Barto, A.G., Williams, R.J.: Reinforcement learning is direct adaptive optimal control. IEEE Control Syst. **12**(2), 19–22 (1992)
41. Vamvoudakis, K.G., Lewis, F.L.: Online actor–critic algorithm to solve the continuous-time infinite horizon optimal control problem. Automatica **46**(5), 878–888 (2010)
42. Watkins, C.J.C.H.: Learning from delayed rewards. PhD thesis, University of Cambridge, England (1989)
43. Whiteson, S., Stone, P.: Evolutionary function approximation for reinforcement learning. J. Mach. Learn. Res. **7**, 877–917 (2006)
44. Wiering, M., Van Otterlo, M.: Reinforcement learning. In: Adaptation, Learning, and Optimization, vol. 12. Springer, Berlin (2012)
45. Williams, R.J., Baird, L.C.: Tight performance bounds on greedy policies based on imperfect value functions. Technical report, Citeseer (1993)
46. Yamaguchi, A., Hyon, S., Ogasawara, T.: Reinforcement learning for balancer embedded humanoid locomotion. In: Humanoid Robots (Humanoids), 2010 10th IEEE-RAS International Conference on, pp. 308–313. IEEE (2010)